# Neural Networks

## IN SEARCH OF MEDIA

Timon Beyes, Mercedes Bunz, and
Wendy Hui Kyong Chun, Series Editors

*Action at a Distance*
*Archives*
*Boundary Images*
*Communication*
*Digital Energetics*
*Machine*
*Markets*
*Media and Management*
*Neural Networks*
*Organize*
*Pattern Discrimination*
*Really Fake*
*Remain*
*Technopharmacology*
*Undoing Networks*

# Neural Networks

**Ranjodh Singh Dhaliwal, Théo Lepage-Richer, and Lucy Suchman**

**IN SEARCH OF MEDIA**

University of Minnesota Press
Minneapolis
London

meson press

In Search of Media is a collaboration between the University of Minnesota Press and meson press, an open access publisher, https://meson.press



Published by the University of Minnesota Press, 2024

111 Third Avenue South, Suite 290
Minneapolis, MN 55401–2520
https://www.upress.umn.edu

in collaboration with
meson press
Salzstrasse 1
21335 Lüneburg, Germany
https://meson.press

ISBN 978-1-5179-1669-5 (pb)

A Cataloging-in-Publication record for this book is available from the Library of Congress.

# Contents

# Series Foreword

"Media determine our situation," Friedrich Kittler infamously wrote in his Introduction to *Gramophone, Film, Typewriter.* Although this dictum is certainly extreme—and media archaeology has been critiqued for being overly dramatic and focused on technological developments—it propels us to keep thinking about media as setting the terms for which we live, socialize, communicate, organize, do scholarship, et cetera. After all, as Kittler continued in his opening statement almost thirty years ago, our situation, "in spite or because" of media, "deserves a description." What, then, are the terms—the limits, the conditions, the periods, the relations, the phrases—of media? And, what is the relationship between these terms and determination? This book series, *In Search of Media,* answers these questions by investigating the often elliptical "terms of media" under which users operate. That is, rather than produce a series of explanatory keyword-based texts to describe media practices, the goal is to understand the conditions (the "terms") under which media is produced, as well as the ways in which media impacts and changes these terms.

Clearly, the rise of search engines has fostered the proliferation and predominance of keywords and terms. At the same time, it has changed the very nature of keywords, since now any word and pattern can become "key." Even further, it has transformed the very process of learning, since search presumes that, (a) with the right phrase, any question can be answered and (b) that the answers lie within the database. The truth, in other words, is "in there." The impact of search/media on knowledge, however, goes beyond search engines. Increasingly, disciplines—from sociology to economics, from the arts to literature—are in search of media as a way to revitalize their methods and objects of study. Our current media situation therefore seems to imply a new term, understood as temporal shifts of mediatic conditioning. Most broadly, then, this series asks: What are the terms or conditions of knowledge itself?

To answer this question, each book features interventions by two (or more) authors, whose approach to a term—to begin with: *communication, pattern discrimination, markets, remain, machine, archives, organize, action at a distance, undoing networks*—diverges and converges in surprising ways. By pairing up scholars from North America and Europe, this series also advances media theory by obviating the proverbial "ten year gap" that exists across language barriers due to the vagaries of translation and local academic customs and in order to provoke new descriptions, prescriptions, and hypotheses—to rethink and reimagine what media can and must do.

## Introduction

# Rendering the Neural Network

**Ranjodh Singh Dhaliwal, Théo Lepage-Richer, and Lucy Suchman**

> "Nature" still serves a foundational role for technoscience, but as a source of certainty and legitimacy for the designed and engineered which, as specifically sited cultural historical enterprises, are rhetorically naturalized.
>
> —Donna Haraway, *Modest–Witness@Second–Millennium*

This book is an exploration of the conjuncture of nature and artifice enacted in the figure of the neural network. In that project it joins a rich body of scholarship devoted to tracing the genealogies through which the biological and the technological have been variously constituted as opposites and as models for each other. As Haraway observes, however much the designed and engineered have achieved ascendancy within the modernist project, their touchstone remains the invocation of nature as their foundational referent. The case of the neural network is no exception, as an organic entity located in the body, and more specifically the brain, is linked to an iconic artifact associated with the manufacture of connections. As the neuron delineates an entity separable from its constitutive relations, the network restores those relations in the form of a generalizable structure. Through these paired moves the neural network is put to work in the service of a more longstanding figure, that of the universal/unmarked human (Wynter 2003).

Disassembling the trope of the neural network enables its recontextualization in specific technoscientific histories, imaginaries, and material practices. The cases that we consider make evident that the neural network serves not only as a consistent referent for the biological and cognitive sciences but also that, like all technoscientific objects, its nature is not fixed nor its futures determined. Our common methodological strategy is to find ways of making the shape-shifting of the neural network evident through a study of key moments in which it is figured and made to work. That in turn reveals how the neuron, the network, and their constitutive relations are contingent on the wider projects—most obviously those of neuroscience and computational engineering, but also the larger sociocultural and politico-economic projects of the world—in which their agencies are enacted and enrolled.

This methodological strategy does not require, or even prefer, either a critical or an appreciative approach determined in advance. Our attitude toward these ever-shifting objects and characters is that of ambivalence, where the scientific and technological worlds enlivened by the many ideas around neural networks are, on the one hand, evidently crucial for their respective social and professional spheres but, on the other hand, also involved in several well-critiqued epistemic problems. In such a scenario, what we advocate is a careful investigation of the contents, characters, and operations in each scene. It is the neural networks that we study that led us to ambivalence, for in their own emergence and sustenance, they stabilize several worlds—technological and scientific but also social, political, cultural, and financial—that do not always align perfectly with the internal arrangement of either neurons or networks. In other words, ambivalence is, in some ways, offered to us by the neural networks themselves. As media theorist Patrick Jagoda points out in his study of network aesthetics (and it is not coincidental that networks lead Jagoda to such an ambivalent position, much as they lead us to ambivalence here), ambivalence is "a mode of extreme presence" that suggests "a crucial critical position from which to think within an uncertain present that is also ongoing" (2016, 224) instead of either embracing or dismissing the object at hand. Ambivalence is an "extreme opting in" that "need not be reduced to naïve complicity or the hyperbolic extremism of strategies such as accelerationism" (225). For us, this ambivalence is important for articulating better understandings of historical and social operations but also for improving critical positions. In the words of Jagoda, such ambivalence involves "a process of slowing down and learning to inhabit a compromised environment with the discomfort, contradiction, and misalignment it entails" (225). This is a form of what Haraway (2016) has named "staying with the trouble." Disassembly, as we envision it, requires such a slowing down, one that is filled with care; after all, going faster only breaks things. In contrast, we aim to go slow and articulate the politics of specific configurations. Ambivalence, thus, should not be confused with the lack of a (political) position; if anything, it is an attention to the very conditions that produce such positions in the first place, both for the characters in our stories and for us as investigators.

With this approach, we aim to interrupt the hegemonic project that neural networks have come to stand for, by questioning what they enable once they are recognized as a privileged model for representing things as varied as brains, learning machines, and complex systems of all kinds. Following the demonstration by Kurt Hornik and his colleagues that "multi-output multilayer feedforward networks" (Hornik, Stinchcombe, and White 1989, 363) can approximate any functions, neural networks have come to be recognized as some sort of universal machine capable of capturing the essential qualities of any system, with little-to-no attention to how these same systems are transformed once they are represented as networks of simple units. Not only does the conceptualization of neural networks as universal machines obscure how it is their malleability—and not their universality—that allows for their

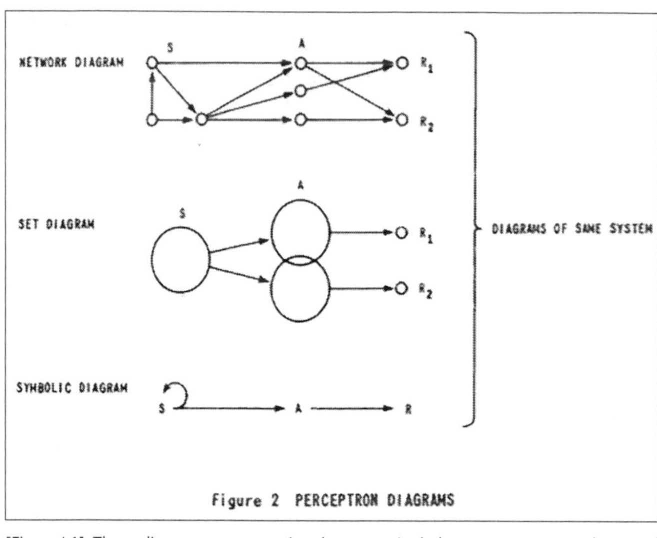


[Figure I.1]. Three diagrams representing the same single-layer, perceptron-style neural network. Frank Rosenblatt, *Principles of Neurodynamics: Perceptrons and the Theory of Brain Mechanisms,* Report No. 1196-G-8, Cornell Aeronautical Laboratory, 1961, 86.

strategic deployment across contexts, but it also obfuscates the politics involved in their characterization as signaling the ultimate convergence of human and machine intelligence.

As we suggest throughout this book, there is nothing deterministic in the way neural networks are represented, conceived of, and made to invoke the neural or even biological domain. Up until the 1960s, there was no consensus vis-à-vis how to represent neural networks, as some of their earlier developers themselves recognized (see Figure I.1). In fact, any given neural network could just as well be represented as a series of matrices multiplying one another, a set diagram, or even a decision tree–like model with weighted branches. That neural networks have come to stand as a logico-mathematical, albeit highly operational, abstraction of the brain's fundamental structures is but the product of sustained efforts to naturalize a specific understanding of intelligence by projecting it onto the brain itself. Our aim with this book is thus not so much to question whether neurons can be modeled as probabilistically weighted units—though we do, of course, question that—as to locate and restore the various settings and configurations through which such biological invocations became, and remained, the structuring figures of contemporary conceptions of artificial intelligence.

## Disassembling the Neural Network

For a machine learning model known for its complexity, opacity, and reliance on large-scale computing infrastructures beyond the reach of most, neural networks have gained a surprisingly large amount of popular attention. For many, neural networks have become widely recognized as the sign of some gradual yet steady convergence of human and artificial intelligence. And yet, behind such a speculative prospect stands a set of highly situated, data-centric practices, whose analogy with the operations of the brain raises the question of whether that convergence might be better understood as the starting point of neuro- and computer science's shared history. If machine learning stands more generally for a statistical approach to the extraction of patterns from large amounts of data, then (artificial) neural networks can be defined as a biologically inspired model relying on probabilistically weighted neuron-like units to identify such patterns. Through their rhetorical deployment of neural tropes, neural networks promote a vision of data extraction and pattern recognition as functions constitutive of the brain itself, advancing a form of recursive thinking about computing and machine intelligence that goes beyond mere analogies and metaphors. In that context, what this book asks is not so much whether neurons can be adequately modeled as simple, information-processing units and neural connections as probabilistically weighted ones; instead, the question it raises is how such computationally inclined representations became, and remained, fundamental to contemporary notions of both human and machine intelligence.

For those unfamiliar with the histories and genealogies that this book both builds on and responds to, we begin with the usual narrative surrounding this particular—and peculiar—model. For computer scientists, the main milestones in the history of neural networks generally begin with their initial introduction by Warren McCulloch and Walter Pitts (1943) as a way to model the mind in terms of the interactions among two-state neurons. Given that binary rendering, neural networks were recast into a promising approach to pattern recognition that could be implemented in computer hardware that was increasingly becoming digital (Selfridge 1955; Rosenblatt 1961). After being abandoned by most in the wake of the rejection of that approach by symbolically inclined AI scholars (Minsky and Papert 1969), neural networks were later rediscovered by both cognitive psychologists and computer scientists, who capitalized on recent advances in processing capabilities and hardware to simulate neural networks in software (Rumelhart, Hinton, and Williams 1986; LeCun 1987; Hinton 1990), fueling a so-called AI renaissance. Various versions of this narrative have been put forward in recent landmark publications such as the field-defining book *Deep Learning* (Goodfellow, Bengio, and Courville 2016) and the much-cited article on convolutional neural networks by the field's three leading figures (LeCun, Bengio, and Hinton 2015). But the teleological core of the story remains the same: for neural networks' proponents, all setbacks in this approach's history are simply the product of limited processing capabilities, the exponential growth of which will eventually prove neural networks' validity as a model of intelligence applicable to both humans and machines.

Many scholars have questioned these internalist histories, noting how their use of neural networks as stand-ins for AI more broadly defined does more to police the limits of what comes to be recognized as intelligence than to shed light on what that capacity might be (Mendon-Plasek 2020; Pasquinelli 2017). More than that, critical scholars have highlighted how algorithmic technologies hard-code various assumptions about the ideal organization of labor (Dhaliwal 2022), social relationships (Chun 2021), populations (Bruder and Halpern 2021), states (Lepage-Richer and McKelvey 2022), and even military intelligence gathering (Suchman 2022) in a way that emphasizes the need to attend to the specific settings in which neural networks are developed and implemented. From these various interventions, there emerges a vision of neural networks as an object of study just as unstable and socially rooted as AI is more generally. Like the wider concepts around AI, neural networks took on several meanings and configurations, which are ultimately linked by little but a broader commitment to form/structure. Arguably it can be claimed that it is neural networks' seemingly endless applicability across contexts that constitutes the core of their appeal, more than any of their individual applications. For when read analogically, neural networks seem to proliferate promiscuously, with many having applied them beyond the realm of computation and neuroscience to reconceive of systems as varied as governments (Deutsch 1966), corporations (Beer 1972), and even whole societies (Luhmann 1995) as neural-like networks of social relationships.

In short, the aim of this book is to disassemble, and resituate, neural networks as a way to recover what happens to brains, machines, and even social systems when they come to be understood in such terms. To initiate these reverse-engineering efforts, we begin by attending to the two entities that converge in the model's very name—namely, neurons and networks.

## Neurons in Context

For a structure that is widely accepted as the basis of all cognitive processes and mental states, neurons have a remarkably controversial history. Formally introduced by Spanish histologist Santiago Ramón y Cajal in 1888 based on a study of the development of the brain in bird embryos, neurons were first defined as "fully autonomous physiological units" (1888, 8) in response to the then dominant conceptualization of the brain as devoid of any substructure whatsoever. Spearheaded by Italian physician Camillo Golgi, who had developed the staining techniques used by Ramón y Cajal to formulate his own countertheory, the latter position posited that the structure of the brain was that of a "diffuse nervous network" (Golgi 1881, 8). Captured as objects of study by carefully slicing and staining brain tissues before observing them under light microscopes, neurons and their alternates relied on the exact same tools and methods, differing only in how the structures revealed by such techniques were interpreted and made sense of.

If this story resonates with most historical studies of the controversy between Golgi and Ramón y Cajal, which have documented how that controversy mapped onto the larger interpretative crisis produced by the rise of new visual practices in the context of turn-of-the-century neurophysiology (Brain 2015; Daston and Galison 2007; Tucker 2005), it differs from the narrative that generally characterizes internalist accounts of the debate. For neuro- and computer scientists alike, contemporaneous advances in microscopy generally remain the main factor explaining the resolution of the debate in Ramón y Cajal's favor (e.g., Fan et al. 2020; Porras and Báguena 2019; Shepherd 2015), positioning the rise of the neuron and its primacy as the obvious outcome of early neurophysiological research. And yet, there is little reason to believe that Ramón y Cajal's images and observations were in any way more accurate or precise than Golgi's (Clarke and Jacyna 1987), as it took a few more decades for the first conclusive images of neural gaps to be finally produced (Palay and Palade 1955). If Ramón y Cajal's contemporaries ultimately adopted his theory over Golgi's, it was instead thanks to a wide range of factors including his proximity with the medical practice, his persona as a professionalized (neuro)scientist, his embrace of experimentation, the parascientific apparatuses that he used (as evidenced by Dhaliwal's contribution in this book), the demonstrations he delivered in Europe and the United States, and the international network he built around his theory (see Star 1989). As part of its appeal, the neuron's introduction then promoted a new way of producing knowledge about life by fixating (dead) biological material, providing a promising alternative to the more philosophical inclinations that animated the work of nineteenth-century vitalists and anatomists alike.

Neurons, in sum, were from the very beginning intimately linked to the apparatus that allowed their rendering as discrete, recognizable units, even though that apparatus might turn out to be more expansive, and less visual in nature, than generally assumed. Notably, the neuron promoted the "histological techniques of killing and solidifying tissues" as privileged means to "delineat[e] the complex structures of organisms" (Landecker 2010, 35). In continuity with previous studies of histology's adoption as a recognized scientific practice (e.g., Hopwood 1999; Jacyna 2001), it is worth asking how the careful and standardized manipulation of dead, human tissues came to be recognized as the ultimate source of explanation for complex living structures. In addition to contributing to a broader transformation of what "counted as an 'explanation' of biological development in individual organisms" (Keller 2003, 3), the neuron's introduction and adoption undermined the explanatory power of other approaches and techniques, notably by decentering the production of knowledge about life from living matter. Furthermore, upon properly inserting the neuroscientific history into the computer-scientific one, one notices the irony in that the history of neurons perhaps involves a disavowal of the nervous "network" at one of its originary scenes, only for that concept to nevertheless end up being associated with the neurocognitive in computation much later, as if the neurons were always already networked for the computational frameworks. Anticipating how neural networks came to define intelligence by its computational, as opposed to biological, manifestations, in this book we explore the broader ramifications of the neural apparatus as a way to recover how neurons came to stand for the universal structure they are now taken to compose, at the expense of the many other biological systems, lived experiences, and embodied practices that have been relegated to the realm of the nongeneralizable.

## Networks in Context

Unlike the neuron, the notion of networks has long been the object of sustained scholarly attention in media studies and kindred fields. The historical literature traces the concept of networks back to the late 1840s, when the development of the telegraph provided a new way to think of communication as a decentralized process, but also of political and economic order as binding together previously distinct entities and agents (Carey 1989). At the same time, the figure of the network also emerged as a promising framework to conceive of the human body and, more specifically, the human brain, as a network-like communication system. While Timothy Lenoir (1986) documents how telegraphic and related technological metaphors shaped the development of new scientific instruments that embodied such views, Laura Otis (2001) highlights how physiological studies inspired communication engineers to embrace a network approach. Born at the intersection of communication engineering and physiology, the notion of network from its first articulation linked the human body with technological systems of all kinds, anticipating their further convergence with the introduction of neural networks.

Through these genealogical connections, the network also emerged as a powerful concept to theorize how communication technologies structure the production and circulation of knowledge, as well as the means through which power and control are exercised. Throughout his oeuvre, Friedrich Kittler notably developed "discourse network" as a concept not only to theorize how "physical, technological, discursive, and social systems [. . .] provide epistemic snapshots of a culture's administration of power and knowledge" (1999, xxxiii–xxiv) but also to posit a break between the patterns of communication of the early and late modern periods. While applying this concept across time periods, Kittler (1990) distinguished the discourse networks of the 1800s onward by their break from previous hierarchical systems models, in which the circulation of information followed structures of social stratification. In contrast,

Kittler describes these latter discourse networks as relying on, and linking, an ever-greater number of objects, communication practices, and inscription methods, thus resisting any form of singular characterization. In addition to theorizing how communication technologies and modes of inscription inform social, cultural, and political systems, such discourse networks marked a moment when knowledge and power themselves became dispersed across various subsystems, which could only be meaningfully analyzed through the study of their interactions. The notion of networks thus provided for Kittler a new way to understand complex systems of knowledge and power by providing a new source of explanation for their relations and dynamics.

Today, networks of all kinds have become objects of regular study and analysis, with scholars across fields both expanding on, and nuancing, Kittler's initial provocation. In this series alone, networks have been alternately historicized (Brunton 2019), undone (Stäheli 2020), and built (Juhasz 2021), in addition to being deployed to account for objects and phenomena as diverse as social platforms (Stiegler 2019), manufacturing (Steinberg 2021), and smart devices (Neves 2022). From all of these accounts emerges a view of networks as not so much a liberatory alternative to other models of power, state, and organization, as they are sometimes made to be, but as a managerial approach in which control is exercised at the level of organization itself, by delimiting the type of relationships that those who inhabit it can develop and maintain (see Galloway and Thacker 2007; Halpern and Mitchell 2022). If networks seem to so readily capture the complexity of the current world, it might then very well be because they performatively intervene upon it, as Wendy Chun (2018) notes, enforcing the worldview they supposedly represent by actively treating the objects and subjects they encompass as simple nodes to be connected and managed. Building upon these various interventions, we explore here what happens when networks are deployed not only to theorize complex systems but also to obfuscate the distinctions among them, be they brains, computing infrastructures, or simulated networks of neuron-like units.

The case studies considered in this book all establish how neural networks shapeshift, how the strategic malleability performed by neural networks is both a formal and a historico-social feature. The ability to change shape is also here accompanied by the ability to move. Precisely because of their status as a boundary object (see Star and Griesemer 1989), their location between nature, science, and technology, and their condition as both physiologized technology and technologized physiology, neural networks prove to be very portable. Science studies has extensively considered how "chains of translation" between field sites and laboratories work through the creation of immutable mobiles, ordering devices through which materials taken from one place become evidentiary support for claims made in others (see, for instance, Latour 1986, 1999). Media and cultural studies scholars have discussed this question with a focus on discourses: theorist Edward Said (1983) has shown the way in which "theory" moves across spaces and times, noting that there is often an inherent disconnect between the historical and social conditions that produce a certain idea and the ways in which academic (and popular) discourse takes up and transforms it. There is always, in this account, a slip between the cup (what makes neurons and networks work at any time and place) and the lip (the readers, the critics, the scientists, the technologists, anyone who cites, and so on). In sum, there are at least two kinds of instabilities that we encounter when we disassemble: there is the historical (or temporal) instability—that is, neurons and networks never remain static across time, let alone neural networks as a concept—and then there is the spatial instability—across disciplines and discourses, neurons, networks, and neural networks all change their significations. Together, this all provides us with a fruitful entry point into the disassembly of the neural network, for neither the concept, nor its valences and situations are given. Insofar as neural networks are both objects (of study) and subjects (as concepts) (see Serres 1997), all of the aforementioned travels apply. The work done by "neural networks," in other words, is not only inseparable from the work done by humans and nonhumans to make neural networks work but is also primarily found in the gaps, that is, in the temporal and spatial instabilities of materials and semiotics that make up neural networks per se.

It might be useful, then, to think of neural networks not as being created, discovered, found, generated, or even studied. Rather, it may be fruitful to understand neural networks as being rendered (see Dhaliwal 2021; Lynch 1985; Myers 2015). Notions of rendering offer an alternative to ideas about re-presentation. What we do here in this volume—and in some ways what our stories show being done—is not so much a representation of historical events or situational circumstances but a rendering thereof. Avoiding objectivist connotations of representation, rendering helps us to unearth the contingency and the instability inherent in processes of techno-scientific inquiry and their relationships with politico-economic and sociocultural domains. Neurons, networks, and neural networks, when disassembled, need to be rendered legible for the scholars working on them and the readers reading about them a rendering process that is itself an extension of prior forms of rendering that make the historical and situational neural networks tick. Each act and each product of rendering partially embraces a story's contingency and refutes its givenness. (Re)producing data (*données,* givens), after all, is not always the same as rendering (articulating) them.

## Rendering 1: Neural Media (Théo Lepage-Richer)

We begin this book with an overview of neural networks' main Iterations in the context of nineteenth-century histology, wartime psychiatry, and late twentieth-century computer science. Taking a media historical approach, which focuses on the broader conditions surrounding neural networks' adoption as a solution to the problem of intelligence at different points in time, this chapter documents neural networks' historical role in mediating the attribution of intelligence across humans and machines. While they are currently known as a biologically inspired, statistical approach to machine intelligence, neural networks were first introduced as a neuroanatomical approach and later a psychiatric model, both of which actively enforced historical conceptions of racial and pathological difference. Materialized through practices as varied as silver staining and colonial health policies, brain lacerations and electroshock therapy, and ultimately organizational reforms and the software implementation of simulated neurons, neural networks were directly inscribed onto certain bodies to produce a recognition of who—or what—qualifies as an intelligent subject and who doesn't. If anything, what unified neural networks' various iterations was a sustained experimental commitment that systematically involved organizing, managing, and disciplining human bodies while devaluing the practical, local, and contextual knowledge they hold. By recovering this broader history, Lepage-Richer proposes to understand neurality itself as a mediating figure that, throughout the twentieth century, conflated communication with structure, meaning with medium, and cognition with neurophysiology. Building on this, he theorizes a broader class of media epitomized by neural networks, which he defines by the way they communicate a highly technologized view of both biology and neurology while promoting the principles they embody—atomization, interdependence, mediation—as universal ones.

## Rendering 2: On Parascientific Mediations: Science Fictions, Educational Platforms, and Other Substrates That Think Neural Networks (Ranjodh Singh Dhaliwal)

We then turn to the environs of technoscientific inquiries into neural networks to look at where thinking about neural networks happens when research is off duty. Beyond, and often in addition to, the usually understood scientific and technological apparatuses and their inscriptions—the notes taken in the lab, the code written on computers, the research articles sent for review, the machines and experimental systems used for research—Dhaliwal finds another category of substrates that help scientists think, iterate, and make sense of neural networks. Locating these parascientific media in odd corners such as lab lit science fiction and self-help advice books written by scientists alongside online educational-course platforms and forms of presentational rhetoric, he finds the history of neural networks not only littered with, but also sutured by, substrates that often function as cognitive sandboxes and social consolidators, helping selectively educate and engineer public opinion about the technoscientific neural. Such parascientific media appear to be critical in thinking through technoscientific concepts but are not usually granted the starring role in conventional stories of scientific controversies or their stabilizations. Here, in two historical slices carved from the story of neural networks—one pertaining to neural debates in biology around the turn of twentieth century and the other marking the connectionist-symbolist artificial intelligence debates in computer science around 2010s—technoscientific controversies, and their outcomes, are shown to be rhetorically stabilized in discourse with the help of parascientific media, which especially help build publics after a scientific controversy has been settled. Several scientific scenes, but specifically ones that involve neural networks, find themselves in need of more space to think, and parascientific media provide one location where the neural network essentially escapes the properly technoscientific setting by splaying itself leisurely while we think—and convince ourselves as much as others of how we think—about thinking.

## Rendering 3: The Neural Network at Its Limits (Lucy Suchman)

We close with a diffractive reading of the transdisciplinary travels of the neuron through the perspicuous cases of two actors currently at work in the fields of computational neural networks and neuroscience. A diffractive reading, following Haraway (1997, 16) and Barad (2007, 71), takes the generation of interference patterns as a method for articulating differences within phenomena; in this case, within technoscientific fields engaged with the neuron as a research object. By bringing computational neural networks researcher Geoffrey Hinton and feminist neuroscientist Gillian Einstein into virtual dialogue, the more specific focus of this chapter is on what we can learn by foregrounding different responses to encounters with the limits of a technoscientific project for which the polysemous neuron is a focal object. While the analogy of brains and computers informs thinking across the biological and computational sciences, it also breaks. Of interest here are ways in which the analogy breaks differently for feminist neuroscientist Gillian Einstein (who follows her research problem from neurons to sexed/gendered bodies and their worlds) than it does for neural networks researcher Geoffrey Hinton (who is committed to sustaining the analogy, even as he acknowledges its limits). Suchman argues that an investment in cognitivism—a theory of intelligence based in a correspondence between mental representations formed in the brain/mind, and a world taken to stand outside of it—and in renderings of cognition as computation sustains the closed-world logics of laboratory computational sciences as well as their methods. In contrast, a commitment to embodied persons in relation entails following neuronal connections across received boundaries between brains, bodies, and worlds. The limits of the neural network in this onto-epistemology comprise generative openings for theoretical and methodological transformation.

## References


Barad, Karen. 2007. *Meeting the Universe Halfway: Quantum Physics and the Entanglement of Matter and Meaning.* Durham, N.C.: Duke University Press.

Beer, Stafford. 1972. *Brain of the Firm: A Development in Management Cybernetics.* Freiburg im Breisgau, Germany: Herder and Herder.

Brain, Robert. 2015. *The Pulse of Modernism: Physiological Aesthetics in Fin-de-Siècle Europe.* Seattle: University of Washington Press.

Bruder, Johannes, and Orit Halpern. 2021. "Optimal Brain Damage: Theorizing Our Nervous Present." *Culture Machine* 20: 1–25.

Brunton, Finn. 2019. "Hello from Earth." In *Communication* by Paula Bialski, Finn

Brunton, and Mercedes Bunz, 1–50. Minneapolis: meson press and University of Minnesota Press.

Carey, James. 1989. *Communication as Culture: Essays on Media and Society.* Boston, Mass.: Unwin Hyman.

Chun, Wendy Hui Kyong. 2018. "Queerying Homophily." In *Pattern Discrimination* by Clemens Apprich, Wendy Hui Kyong Chun, Florian Cramer, and Hito Steyerl, 59–98. Minneapolis: meson press and University of Minnesota Press.

Chun, Wendy Hui Kyong. 2021. *Discriminating Data: Correlation, Neighborhoods, and the New Politics of Recognition.* Cambridge, Mass.: MIT Press.

Clarke, Edwin, and L.S. Jacyna. 1987. *Nineteenth-Century Origins of Neuroscientific Concepts.* Berkeley: University of California Press.

Daston, Lorraine, and Peter Galison. 2007. *Objectivity.* New York: Verso.

Deutsch, Karl. 1966. *The Nerves of Government: Models of Political Communication and Control.* New York: Free Press.

Dhaliwal, Ranjodh. 2021. "Rendering the Computer: A Political Diagrammatology of Technology." PhD diss. University of California, Davis. https://escholarship.org/uc/item/13k7k3zh.

Dhaliwal, Ranjodh. 2022. "The Cyber-Homunculus: On Race and Labor in Plans for Computation." *Configurations* 30, no. 4: 377–409.

Fan, Jingtao, Lu Fang, Jiamin Wu, Yuchen Guo, and Qionghai Dai. 2020. "From Brain Science to Artificial Intelligence." *Engineering* 6: 248–52.

Galloway, Alexander, and Eugene Thacker. 2007. *The Exploit.* Minneapolis: University of Minnesota Press.

Golgi, Camillo. 1881. *Sulla origine centrale dei nervi.* Naples, Italy: Enrico Detken Editore.

Goodfellow, Ian, Yoshua Bengio, and Aaron Courville. 2016. *Deep Learning.* Cambridge, Mass.: MIT Press.

Halpern, Orit, and Robert Mitchell. 2022. *The Smartness Mandate.* Cambridge, Mass.: MIT Press.

Haraway, Donna. 1997. *Modest−Witness@Second−Millennium. Femaleman−Meets−Oncomouse: Feminism and Technoscience.* New York: Routledge.

Haraway, Donna. 2016. *Staying with the Trouble: Making Kin in the Chthulucene.* Durham, N.C.: Duke University Press.

Hinton, Geoffrey. 1990. "Connectionist Learning Procedures." *Artificial Intelligence* 40: 185–234.

Hopwood, Nick. 1999. "'Giving Body' to Embryos: Modeling, Mechanism, and the Microtome in Late Nineteenth-Century Anatomy." *Isis* 90: 462–96.

Hornik, Kurt, Maxwell Stinchcombe, and Halbert White. 1989. "Multilayer Feedforward Networks Are Universal Approximators." *Neural Networks* 2: 359–66.

Jacyna, L. Stephen. 2001. "'A Host of Experienced Microscopists': The Establishment of Histology in Nineteenth-Century Edinburgh." *Bulletin of the History of Medicine* 75: 225–53.

Jagoda, Patrick. 2016. *Network Aesthetics.* Chicago: University of Chicago Press.

Juhasz, Alexandra. 2021. "Who Contrives the Moment? On Cyberfeminist Dating." In *Really Fake* by Alexandra Juhasz, Ganaele Langlois, and Nishant Shah, 10–51. Minneapolis: meson press and University of Minnesota Press.

Keller, Evelyn Fox. 2003. *Making Sense of Life: Explaining Biological Development with Models, Metaphors, and Machines.* Cambridge, Mass.: Harvard University Press.

Kittler, Friedrich. 1990. *Discourse Networks 1800/1900.* Stanford, Calif.: Stanford University Press.

Kittler, Friedrich. 1999. *Gramophone, Film, Typewriter.* Stanford, Calif.: Stanford University Press.

Landecker, Hannah. 2010. *Culturing Life: How Cells Became Technologies.* Cambridge, Mass.: Harvard University Press.

Latour, Bruno. 1986. "Visualization and Cognition." *Knowledge and Society* 6, no. 6: 1–40.

Latour, Bruno. 1999. *Pandora's Hope: Essays on the Reality of Science Studies.* Cambridge, Mass.: Harvard University Press.

LeCun, Yann. 1987. "Modèles connexionistes de l'apprentissage." PhD diss. Université de Paris VI.

LeCun, Yann, Yoshua Bengio, and Geoffrey Hinton. 2015. "Deep Learning." *Nature* 521: 436–44.

Lenoir, Timothy. 1986. "Models and Instruments in the Development of Electrophysiology, 1845–1912." *Historical Studies in the Physical and Biological Sciences* 17, no. 1: 1–54.

Lepage-Richer, Théo, and Fenwick McKelvey. 2022. "States of Computing: On Government Organization and Artificial Intelligence in Canada." *Big Data & Society* 9, no. 2: 1–15.

Luhmann, Niklas. 1995. *Social Systems.* Stanford, Calif.: Stanford University Press.

Lynch, Michael. 1985. *Art and Artifact in Laboratory Science: A Study of Shop Work and Shop Talk in a Research Laboratory.* London: Routledge.

McCulloch, Warren, and Walter Pitts. 1943. "A Logical Calculus of the Ideas Immanent in Nervous Activity." *Bulletin of Mathematical Biophysics* 5: 115–33.

Mendon-Plasek, Aaron. 2020. "Mechanized Significance and Machine Learning: Why It Became Thinkable and Preferable to Teach Machines to Judge the World." In *The Cultural Life of Machine Learning,* ed. Jonathan Roberge and Michael Castelle, 31–78. London: Palgrave Macmillan.

Minsky, Marvin, and Seymour Papert. 1969. *Perceptrons: An Introduction to Computational Geometry.* Cambridge, Mass.: The MIT Press.

Myers, Natasha. 2015. *Rendering Life Molecular: Models, Modelers, and Excitable Matter.* Durham, N.C.: Duke University Press.

Neves, Joshua. 2022. "The Internet of Things and People." In *Technopharmacology* by Joshua Neves, Aleena Chia, Susanna Paasonen, and Ravi Sundaram, 89–119. Minneapolis: meson press and University of Minnesota Press.

Otis, Laura. 2001. *Networking: Communicating with Bodies and Machines in the Nineteenth Century.* Ann Arbor: University of Michigan Press.

Palay, Sanford, and George Palade. 1955. "The Fine Structure of Neurons." *Journal of Biophysical and Biochemical Cytology* 1, no. 1: 69–107.

Pasquinelli, Matteo. 2017. "Machines That Morph Logic: Neural Networks and the Distorted Automation of Intelligence as Statistical Inference." *Glass Bead* 1: 1–17.

Porras, María, and María Báguena. 2019. "The Major Discoveries of Santiago Ramón

Cajal and His Disciples: Consolidated Milestones for the Neuroscience of the 21st Century." *European Journal of Anatomy* 23, no. 1: 5–7.

Ramón y Cajal, Santiago. 1888. "Estructura de los centros nerviosos de las aves." *Revista trimestral de histología normal y patológica* 1: 1–10.

Rosenblatt, Frank. 1961. "Principles of Neurodynamics: Perceptrons and the Theory of Brain Mechanisms." Buffalo, N.Y.: Cornell Aeronautical Laboratory.

Rumelhart, David, Geoffrey Hinton, and James Herbert Williams. 1986. "Learning Internal Representations by Error Propagation." In *Parallel Distributed Processing, Volume 1,* ed. David Rumelhart and James McClelland, 318–62. Cambridge, Mass.: MIT Press.

Said, Edward. 1983. *The World, the Text, and the Critic.* Cambridge, Mass.: Harvard University Press.

Selfridge, Oliver. 1955. "Pattern Recognition and Modern Computers." In *Proceedings of the March 1–3, 1955, Western Joint Computer Conference,* 51–93. New York: Association of Computing Machinery.

Serres, Michel. 1997. *Le parasite.* Paris: Grasset.

Shepherd, Gordon. 2015. *Foundations of the Neuron Doctrine: 25th Anniversary Edition.* Oxford: Oxford University Press.

Stäheli, Urs. 2020. "Undoing Networks." In *Undoing Networks,* by Tero Karppi, Urs Stäheli, Clara Wieghorst, and Lea Zierott, 1–30. Minneapolis: meson press and University of Minnesota Press.

Star, Susan Leigh. 1989. *Regions of the Brain: Brain Research and the Quest for Scientific Certainty.* Stanford, Calif.: Stanford University Press.

Star, Susan Leigh, and James Griesemer. 1989. "Institutional Ecology, 'Translations' and Boundary Objects: Amateurs and Professionals in Berkeley's Museum of Vertebrate Zoology, 1907–39." *Social Studies of Science* 19, no. 3: 387–420.

Steinberg, Marc. 2021. "Management's Mediations: The Case of Toyotism." In *Media and Management.* Minneapolis: University of Minnesota Press.

Stiegler, Bernard. 2019. "For a Neganthropology of Automatic Society." In *Machine.* By Thomas Pringle, Gertrude Koch, and Bernard Stiegler, 25–48. Minneapolis: meson press and University of Minnesota Press.

Suchman, Lucy. 2022. "Imaginaries of Omniscience: Automating Intelligence in the US Department of Defense." *Social Studies of Science* 53, no. 5: 1–26.

Tucker, Jennifer. 2005. *Nature Exposed: Photography as Eyewitness in Victorian Science.* Baltimore: Johns Hopkins University Press.

Wynter, Sylvia. 2003. "Unsettling the Coloniality of Being/Power/Truth/Freedom: Towards the Human, After Man, Its Overrepresentation—An Argument." *CR: The New Centennial Review* 3, no 3: 257–337.

[ 1 ]

# Neural Media

**Théo Lepage-Richer**

"For a long time, the human was something else altogether; it is not so long ago that it became a machine—a calculating one no less."[1] However contemporary this statement might sound, it wasn't formulated by some famous media philosopher or posthumanist scholar. Instead, it is by French sociologist Marcel Mauss (1923, 176–77), who, a century ago, commented on the growing influence of a new conceptualization of intelligence as the product of tightly interconnected webs of calculating units initially called "réseau du soma neuronal" (Ramón y Cajal 1907, 19) and later "neural networks." When introduced by Spanish histologist Santiago Ramón y Cajal at the turn of the century, this model involved reconceiving of the brain as being constituted of discrete nerve cells, which provided a new fundamental unit to locate, measure, and attribute intelligence. At a time when advances in histology showed that the brain of all humans was essentially identical at a microscopic level, neural networks constituted a new, albeit yet-to-be-seen, structure that reinscribed within the brain assumed differences in intellectual potential between Europeans and the populations native to their colonies. Decades before they would be embodied by learning machines and artificial intelligence (AI), it was thus the racialized body of colonial subjects that provided the vehicle for the promotion of new ideas about intelligence and its reducibility to locatable units—a reconceptualization that was as much neurophysiological as sociological, if not altogether racially inclined.

In this chapter, I take as my case studies three such episodes when neural networks were not only deployed as solutions to the problem of intelligence but also framed as such by being directly inscribed onto specific bodies. Now known as a biologically inspired machine learning model,[2] neural networks were first introduced as a neuroanatomical approach to racial difference before being recast as a new way to both diagnose and treat mental illness based on neurophysiology alone. Traversing fields and time periods as varied as nineteenth-century histology, World War II-era psychiatry, and late twentieth-century computer science, neural networks have indexed not only changes in the articulation of intelligence but also the means developed to locate and attribute the latter across humans and machines alike. Despite their assumed coherence through time, neural networks have been materialized through practices as varied as silver staining and colonial health policies, brain lacerations and electroshock therapy, and finally organizational reforms and the programming of software-simulated neurons. If anything, what encompasses these iterations is not so much some unified theory of intelligence as an experimental ethos that systematically involved organizing, managing, and disciplining human bodies while devaluing the practical, local, and contextual knowledge they hold. Turning on its head Lucy Suchman's invitation to attend to the sites where the prospect of "machines-as-agents" is produced (1987, 2), this chapter then returns to three sites where the prospect of agents-as-machines—i.e., that agency and, in that specific case, intelligence necessarily involves machinic qualities—was first articulated and later sustained by denying various bodies their agential potential.

The three episodes in question are the following: the discovery of biological neural networks by Spanish histologist Santiago Ramón y Cajal (1888a), the invention of artificial neural networks by American psychiatrist Warren McCulloch (McCulloch and Pitts 1943), and the software implementation of deep neural networks by British-Canadian computer scientist Geoffrey Hinton (Hinton, McClelland, and Rumelhart 1986). In each of these episodes, neural networks were deployed to stabilize ongoing crises surrounding the definition and operationalization of intelligence, but also introduced enough instability to justify subjecting various bodies and objects to the experimental method. To explore this dialectical relationship between stability and instability in the history of neural networks, I draw on Hans-Jörg Rheinberger's work on experimental systems. Defined by Rheinberger as "vehicles for materializing questions" (1997, 28), experimental systems consist in the fundamental unit of experimentation, where the technical, epistemic, social, and institutional components of laboratory work converge in the form of situated objects (1997, 238). Since its introduction, this concept has been appropriated by media studies scholars (e.g., Jagoda 2020; Milburn 2010) to theorize how media render new situations, settings, and cultural milieus available to experimentation—a move that provides deeper critical purchase on neural networks' mobilization of bodies of all kinds as objects of histological, clinical, and ultimately computational manipulation.

With this piece, I propose to recenter the study of neural networks on the commitment to experimentation they embody, while also theorizing a broader class of media defined by this commitment. This contribution is key, as it provides a more robust framework to conceive of neural networks than their usual characterization as a biologically plausible alternative to symbolically inclined approaches to intelligence. For despite being generally defined in opposition to one another, biologically inspired models like neural networks and symbolic conceptions of AI share similar assumptions vis-à-vis the computational basis of cognition, as Lucy Suchman points out later in this book. The latter might posit that cognition involves computing symbols (e.g., Newell 1980; Pylyshyn 1984) whereas the former that it is the product of serialized computing units (e.g., Hinton, McClelland, and Rumelhart 1986; Rumelhart and McClelland 1987), but they still exhibit a similar understanding of computation as the main currency of cognition. In that context, what neural networks' experimental underpinnings provide is a powerful lens to define this model by its effects on the objects it is applied to. In each of my cases, experimentation was deployed to inscribe the principles embodied by neural networks onto the same bodies whose agency they denied. Throughout the history of neural networks, the brain and its increasingly abstract forms were systematically mobilized as vehicles to communicate new ideas about atomization, interdependency, and mediation. What this emphasis on experimentation thus highlights is the central role of mediation in the development of neurality itself, which Ramón y Cajal, McCulloch, and Hinton alike all invoked to promote their understanding of human biology, psychology, and sociality as reducible to locatable structures available for experimentation.

Building on Clifford Siskin's definition of mediation as "a form that works physically in the world to mediate our efforts to know it" (2016, 1),[3] I thus propose to understand Ramón y Cajal's, McCulloch's, and Hinton's various formulations—and inscriptions—of neural networks as representative of a subset of media I term *neural media.* With it, I aim to theorize the various objects deployed by my protagonists to promote neurality as a form as well as naturalize it as a biological given. For in each of their iterations, not only were neural networks directly inscribed onto certain bodies through various experimental techniques, but they were also deployed to construct these same bodies as definite proofs of the reducibility of complex mental states and faculties to a neurophysiological basis. As such, neural networks epitomized a subset of experimental systems that, throughout the twentieth century, conflated cognition with neurophysiology and, by extension, communication with structure, meaning with medium. Along with neural networks, we could include in this category other artifacts and practices discussed by other contributors to this series, such as neurohacking (Chia 2022), "smart" devices (Neves 2022), and even internet networks (Stäheli 2020). In all these cases, it is the notion of the neural—or of neural-like networks—that is itself the outcome of the gradual abstraction of the body into brains, minds, and ultimately intelligence, which all provide increasingly intricate proxies for phenomena both social and psychological that otherwise evade experimental and technological mediation. With the notion of neural media, it is then the convergence of the human and the machinic into an all-encompassing theory of organization that I aim to capture—one in which the promotion of atomization, interdependence, and mediation as universal principles is predicated on their gradual dislocation from any body whatsoever.

## Modeling the Brain: On Racial Difference and Colonial Expansion in Nineteenth-Century Histology

Neural networks' first iteration in the context of nineteenth-century histology is also their most overlooked one in the historical and critical literature on AI. While regularly referred to by its subsequent champions as the neurophysiological origin of their eponymous models (e.g., McCulloch and Pfeiffer 1949; Rumelhart, McClelland, and Hinton 1986), neural networks' first iteration as a model positing that all mental states and intellectual faculties were reducible to the interactions among discrete cells is generally left unproblematized. But when first introduced, neural networks directly responded to growing anxieties about the biological basis of racial difference—or its lack thereof. At a technoscientific level, their introduction coincided with key advances in microscopy, which, far from illuminating meaningful physiological differences across racial divides, highlighted the indistinguishability of human tissues at a microscopic scale. At a political level, neural networks' formulation and subsequent adoption unfolded alongside the gradual dissolution of many European colonial empires. From Gayatri Spivak to Sylvia Wynter, many have noted how this dissolution translated in the displacement of strict oppositions between Europeans and their colonized "Others" by more subtle "economic, political, and culturalist maneuvers" (Spivak 1999, 172) that mapped these oppositions "onto the range of human hereditary variations and their cultures" (Wynter 1995, 34). At the intersection of both sets of events, neural networks introduced a more flexible model of neurological and social organization, which not only inscribed racial hierarchies into the structure of the brain but also captured colonized populations as privileged objects of experimental manipulation.

When first introduced by Spanish histologist Santiago Ramón y Cajal, who claimed that the brain was made of "fully autonomous physiological units" (1888a, 8), neural networks opposed the then-dominant conception of the brain as constituted of continuous nervous fibres. While initially formulated by Joseph von Gerlach (1872) and Theodor Meynert (1872), the latter position was by then championed by Italian physician Camillo Golgi, who introduced in 1873 a novel staining technique to demonstrate how the brain was but "a diffuse nervous network" (Golgi 1907, 14). Called "black reaction," this method consisted in immersing thin slices of brain tissues in potassium dichromate before soaking them in a solution of silver nitrate until they developed a blackened coloration (see Golgi 1873). Once subjected to this process, brain tissues could be observed with unprecedented clarity, as the demarcations among their most intricate structures became more visible under light microscopy. Interestingly, however, it is by using the exact same technique that Ramón y Cajal formulated his own neuroanatomical model. Relentlessly slicing, staining, scrutinizing, and drawing brain tissues, Ramón y Cajal and Golgi both captured the brain as available to manipulation and observation but ultimately disagreed on the structures made visible by this process. Presented with the same structures, Golgi saw continuity whereas Ramón y Cajal perceived distinct units. With the first definite images of neural independence being produced only decades later (see Palay and Palade 1955), it is then not so much what they saw that guided their respective models as how their models allowed them to render other objects—and subjects—available to similar experimental manipulations as those they subjected the brain to.

On the one hand, it was primarily to provide a clear material basis to the neurological and psychological illnesses he encountered in his work that Golgi first introduced his model. A trained psychiatrist, Golgi developed his famous staining method while working with tissues taken from deceased patients from the Hospital of San Matteo[4] where he worked (see Golgi 1869a, 1869b). While remembered for his extensive studies of key brain structures using his staining technique, Golgi for his part conceived of his technique as a "positive and experimental" alternative to the "predominantly philosophical-speculative tendencies in the study of mental diseases" (Golgi, quoted in Mazzarello 2009, 387). Producing his first silver-stained studies on the brain of a patient having suffered from chorea—then understood as a manic condition—Golgi (1874) introduced his nervous networks as a privileged model to not only map onto the brain otherwise abstract psychological phenomena but also transform the practice of psychiatry into an experimental one targeting the whole nervous system.

On the other hand, it was rather the pursuit of a material explanation for human development that animated Ramón y Cajal's articulation of his neural networks model. Throughout his career, Ramón y Cajal regularly returned to the question of how human life "began unconscious and ended conscious" (1989 [1901], 293), producing extensive studies of human tissues like bone marrow (1887), the nervous system (1888a), and the brain (1888b) to provide answers to that question. But the type of human development he concerned himself with was directly inspired by the vocabulary and frameworks of late nineteenth-century race science. In another formulation of his work's central question, Ramón y Cajal wondered how some humans could be "the slave and plaything of the cosmic forces" while others were "the driver of nature and the autocrat of creation" (1989, 293). Faced with undistinguishable tissues, Ramón y Cajal posited that nothing but neural independence could explain the wide variations in human capacities and faculties he assumed differentiated "the slave and plaything" from "the driver of nature." Like members of a society who, "in their subordination to an organic whole, still enjoy a certain degree of functional autonomy" (1904, 189), neurons constituted for Ramón y

Cajal a reflection of how humankind was constituted of a manifold of closely interacting individuals, which he conceived of as collectively driven by the near-heroic contribution of its most exceptional members (see 1999 [1897], 23).

So the type of social order Ramón y Cajal imagined as the mirror image of neurological organization was indeed reminiscent of the one he himself inhabited, as it posited a highly stratified structure similar to the strict racial hierarchies underpinning the organization of the late Spanish empire. Transitioning to the study of neuroanatomy while deployed as a medical officer during the first Cuban war of independence, Ramón y Cajal interpreted the clear social stratifications that organized life in colonial Cuba as a proof of the fundamental racial qualities embodied by the colony's different classes. At the bottom of the island's hierarchy, Ramón y Cajal identified "the native race," which occupied the most remote parts of the island and mostly partook in subsistence farming, and "the Africans and Mulattos," whose physical strength made them especially well adapted to agricultural labour (1989, 212). Then, he listed the Creoles,[5] i.e., landowners of mixed but mostly Spanish descent, whose exposure to tropical conditions over several generations had turned into "pale hothouse plants living indolently and parasitically" (1989, 212). On top of them, Ramón y Cajal singled out the Cubans—natives of the island with recent and direct European ancestry—who, "confined to the city and engaged in business or professional work" (1989, 212), possessed the necessary intellect and local knowledge to act as mediators between colonial Cuba and peninsular Spain. And supplanting them all, Ramón y Cajal finally located the "white race," i.e., peninsular Spaniards like himself, who, despite their vulnerability to "the enervating effects of the climate," dominated all the other groups by their superior intellectual "vigour" (1989, 212–13). In the image of his later neural networks model, Cuban society stood for Ramón y Cajal as a distributed, albeit hierarchical network of clearly differentiated groups and individuals, whose stratified structure allegedly mirrored that of racial development itself.

Following his return to Spain, Ramón y Cajal transposed this racialized view of human development into a highly hierarchical understanding of animal biology. Rejecting the hypothesis that "the differences between the brain of lower mammals (cat, dog, monkey, etc.) and that of man are purely quantitative" (1989, 471; see also 1897, 132–33), Ramón y Cajal hypothesized the existence of discrete units within the brain whose number mattered less than that of their connections. Promoting his ideas through the hundreds of histological studies he produced, Ramón y Cajal emphasized how, to the same extent that "the whale's or elephant's big brain" did not correlate with "greater intelligence," the larger brain of certain human groups signaled "inferior intelligence if not altogether stupidity" because of "the imperfect connections among its neurons" their greater number implied (1894, 467). By providing a material explanation for "the great differences both quantitative and qualitative in mental capabilities both across species and within a single one" (1894, 468), neural networks thus projected at a neurophysiological level the racial divides Ramón y Cajal perceived across human populations.

But in addition to providing a biological basis to the racial hierarchies animating the colonial worldview of nineteenth-century Spain, Ramón y Cajal's neural networks also introduced a certain level of indeterminacy that enabled their experimental ethos to expand beyond the realm of histology. For at the same time as Ramón y Cajal equated the intellectual superiority of the "white race" with its capacity to reshape environments otherwise "uninhabitable for the European" (1989, 205), he also singled out this latter endeavor as a necessity for the survival of the European race. More than local insurgencies and anticolonial sentiments, it is the diseases native to Europe's colonies that Ramón y Cajal recognized as the main threat to human development. More than an obstacle to colonization, these diseases consisted for Ramón y Cajal in a direct threat to the boundaries otherwise assumed between European and colonial bodies by indiscriminately jumping from one body to the next. Reduced to carriers for the proliferation of such forms of "parasitic life," which "swept over our couches, raided our provisions, and enveloped us from every side" (1989, 217), native populations stood as little but stand-ins for otherwise invisible pathogens, with Ramón y Cajal calling for their sustained control and management to prevent these pathogens from spreading across racial divides. In addition to the brain, it was thus also the many populations native to Spain's colonies that Ramón y Cajal captured as objects of intervention through his model by notably recasting colonization as a sanitary endeavor with clear racial undertones.

Ramón y Cajal's recognition of colonized populations as proxies for otherwise invisible diseases is especially key, given his role on a major public commission specifically convened in 1910 to improve the sanitary conditions in the Gulf of Guinea. At this time, Spain's overseas territories in Africa were in fact its only colonial possessions, Cuba and the Philippines having both acquired their independence in 1898. In the resulting report, Ramón y Cajal stressed that the economic exploitation of Spain's last colony required sanitizing what essentially amounted to "a paradise for all pathogens" (1910, 9). As a solution, Ramón y Cajal invoked the creation of medical brigades, which would perform through medicine what their military counterparts performed through policing. Analogous in their function to the latter's search for, and elimination of, "hidden enemies," Ramón y Cajal imagined these brigades as composed of scientists mandated to "study prevailing diseases in the colonies; identify their cause in light of the latest bacteriological methods; and propose the necessary prophylactic measures to eradicate them" (1910, 11). To justify this policy, the report featured various experimental studies on human tissues infected by some of the parasites endemic to the region. Tellingly, none of these studies documented the minute neurophysiological differences between Europeans and non-Europeans Ramón y Cajal had previously theorized. Instead, they almost all focused on the parasite responsible for trypanosomiasis, a disease with severe neurological symptoms, in a way that constructed the native body and more specifically brain as a privileged site of parasitic contamination (e.g., Pittaluga 1910). In the name of documenting local diseases and treating native populations, these studies of infected neural networks characterized native populations as bearers of a profound pathological difference necessitating the sustained intervention of a foreign power like Spain. More than a neuroanatomical model, Ramón y Cajal's neural networks thus also stood as a framework to study, identify, and eradicate various diseases and parasites, in a way that reaffirmed assumptions about racial difference under a medical pretense.

Far from being the straightforward product of empirical observations, neural networks then first stood as complex experimental constructs produced through the careful manipulation of human tissues and colonized populations alike. By guiding practices as diverse as silver staining and colonial health policies, neural networks provided a highly operational device to maintain uneven power relations at both a micro- and a macroscopic scale. As such, they actively participated in what others have previously described as "a distinctive late-colonial mode of population management" (W. Anderson 2006, 4) defined by its reinscription of enduring conceptions of racial difference into a medical framework (see Kramer 2006; Neill 2012). Part experimental, part speculative, neural networks not only reflected a wide range of anxieties vis-à-vis human difference, contagion, and colonial dissolution but also initiated a broader reconceptualization of biological life in terms of fundamental structures that could be intervened on. Therefore, Ramón y Cajal's neural networks were not so much discovered as directly inscribed onto brain tissues through staining techniques, mediating how certain bodies were constructed as both racialized and medicalized Others. At the same time as they were deployed to settle the problem of how best to understand the brain and its structures, neural networks then also provided a solution to the problem of how to both justify and renew Spain's colonial project by providing a biological and medical basis for the interdependency between Europe and its colonies.

## Embodying the Mind: On Signals and Electroshocks in Wartime Psychiatry

Neural networks' second iteration in the context of mid-century psychiatry is the usual starting point of most historical and technical accounts of the machine learning model of the same name. While cognitive and computer scientists attribute to that period's work the original insight that "neural-like networks [can] compute" (Rumelhart and Zipser 1986, 152), historians characterize this iteration as a first attempt at providing a mathematical distillation of brain activities (e.g., Dupuy 1994; Kay 2001). But what both sets of accounts overlook is the clear experimental quality of neural networks' second iteration. First studied through techniques like brain lacerations and strychnine neurography before guiding clinical trials exploring the use of electroshock therapy, neural networks' second iteration indexed an experimental and neurophysiological shift in both the practice and theory of psychiatry. Introduced as an alternative to psychoanalysis and behaviorism, which then dominated psychiatric institutions in the United States, neural networks aimed to provide a biological basis for the diagnosis and treatment of mental illness. Designed with the assumption that certain conditions were beyond the reach of psychosomatic lines of treatment, neural networks singled out certain minds as the bearers of a profound pathological difference that could only be treated through somatic interventions. As such, years before they became the flag bearer of the new "science of signals [. . .] called cybernetics" (McCulloch and Pfeiffer 1949, 373), neural networks were first upheld as a model of the mind through their deployment in theorizing some of the defining conditions of twentieth-century psychiatry.

Nearly fifteen years before he formally introduced neural networks as a logico-mathematical model of the mind applicable to both brains and computers, Warren McCulloch entered the psychiatric practice at a time of great turmoil. While psychoanalysis and behaviorism constituted the de facto disciplinary lenses for most professional psychiatrists, neurology and neurophysiology were gradually making their way into psychiatric institutions as promising frameworks to treat conditions for which no effective treatment was yet available (Star 1989; Weinstein 2013). Beginning his career at the Bellevue Hospital in New York in 1928, McCulloch quickly identified with the latter neuropsychiatric movement while working on the impact of brain injuries on the development of neurological and psychological disorders. Alongside his clinical practice, and in collaboration with neurologist Robert Kennedy, McCulloch conducted his first neuropsychiatric experiments by carefully lacerating the brains of cats to study these injuries' impact on mental processes. It is only after 1934, however, that McCulloch's interest in the physical basis of mental illness translated into a coherent experimental agenda centered on the study of neural activities. Appointed at Yale University's Laboratory of Neurophysiology, McCulloch collaborated with physiologist Joannes Dusser de Barenne to devise experiments whose aim was to map the functional relationship across brain regions. Using monkeys as their experimental subjects, Dusser de Barenne and McCulloch developed a method called "strychnine neurography" (1939, 620), which consisted in measuring how variations in one brain region altered electrophysiological readings in others. Injecting a small amount of strychnine—a powerful neurotoxin with highly stimulant properties—directly into their experimental subjects' brain, they produced complex diagrams visualizing the ramifications of different neural pathways when exposed to strychnine (see Dusser de Barenne, Garol, and McCulloch 1941).

In his initial studies of neural activities, McCulloch thus rendered neural pathways available as objects of experimentation by directly inscribing them onto the brain of his experimental subjects through a combination of brain lacerations and targeted poisoning. But despite having no immediate clinical application, McCulloch's experiments nevertheless aimed to develop new ways to not only understand but also treat mental illness. Pursuing these experiments with the intent "of learning enough physiology of man to understand how brains work" (McCulloch 1974, 30), McCulloch strived to identify proxies for mental states and phenomena that otherwise seemed to be beyond the reach of psychiatry. In line with the then-ongoing reconceptualization of the mind as a privileged target of medical intervention, McCulloch adopted the brains of cats and monkeys as proxies not so much for the human brain as for the diseased mind (Grob 1983, 108–43; see also Abraham 2016)—a new object of concern that preoccupied both McCulloch and the many institutions built throughout the interwar period to accommodate the United States' growing mental health population. As he wrote at the time, McCulloch upheld the view that "we will get nowhere with crazy people until we can understand brains in such physio-chemical terms as we use when thinking of kidneys" (1941, 1) and thus conceived of his complex diagrams of neural pathways as conveying the promise of transforming psychiatry into an empirical and, ultimately, experimental practice.

It is at this time that McCulloch's experimental and clinical practices converged, when he took on the position of professor of psychiatry at the Illinois Neuropsychiatric Institute (INI)—one of the few institutions in the United States where fundamental brain research and the clinical treatment of mental health patients coexisted within the same walls. There, while continuing to refine his modeling of neural pathways (e.g., Bonin, Garol, and McCulloch 1942), McCulloch focused on reproducing in experimental subjects some of the conditions he encountered at the INI. Attending primarily to neurosis and psychosis, McCulloch worked on identifying the neural pathways responsible for these pathologies, with the stated intention of developing new lines of treatment for patients diagnosed with them. It is also at that time that McCulloch met his future collaborator Walter Pitts, then a sixteen-year-old self-taught runaway with a talent for formal logic. Recognizing Pitts's work on formal systems as analogous to his research on the brain, McCulloch took him in and raised him alongside of his own children. Under Pitts's guidance, McCulloch's neural pathways were for the first time abstracted from their neurophysiological basis and transposed into a framework inspired by that of electrical engineering. Reinterpreting McCulloch's "neural pathways" into "neuron networks," Pitts proposed to model the connections among neurons as "a simple circuit" (1942a, 121) in which two-valued units are combined together to reach "steady-state equilibria" (1942b, 169).

It is from this mix of clinical, laboratory, and theoretical work that McCulloch and Pitts formally introduced their "neural nets" model in 1943, in a paper densely titled "A Logical Calculus of the Ideas Immanent in Nervous Activity." In it, McCulloch and Pitts did away with the actual mechanisms through which neurons enter excitatory or inhibitory states and instead focused on the logical relationships they embody. Using Boolean algebra—a type of mathematical logic positing that all logical propositions can be distilled into symbols representing discrete states and types of relation—as its system of notation, the piece proposed to reduce neurons to discrete values like on and off or 1 and 0, while representing the relationships among them as simple functions like conjunction, disjunction, and negation (1943, 115–20). Through the combination of such simple states and functions, McCulloch and Pitts claimed that their networks of neurons could embody complex representations when deployed at a certain scale, including any mental state the brain was capable of upholding.

But however abstract and "unnecessarily complicated" (Fitch 1944, 49) their new model was made to be by the piece's early readers, McCulloch and Pitts primarily conceived of their model as an objective lens to "diagnose [. . .] the organically diseased" (1943, 132). While later characterized by cyberneticists as a model of feedback in biological systems (Wiener 1945) and even the basis for a general theory of automata (Neumann 1951), neural networks were first introduced as a privileged framework to define ambiguously differentiated conditions such as hysteria, neurosis, and psychosis. At its core, what this model aimed to demonstrate was how "diseased mentality can be understood without loss of scope or rigour, in the scientific terms of neurophysiology" (1943,

132) with no need for patients' own accounts of their condition and experience. As such, in their second formulation in the context of wartime psychiatry, neural networks were validated as a model of the mind by the way they raised the prospect of reducing mental disorders that otherwise resisted explanation and treatment to locatable brain structures that could be intervened on.

In the following years, neural networks' reconceptualization of mental disorders as physical conditions that could be treated through the direct manipulation of the brain was notably upheld through their application to the treatment of schizophrenia. The choice of this disorder by McCulloch was far from neutral, given both its privileged status in the history of modern psychiatry and its broader philosophical resonance. If the likes of Bertrand Russell and Alfred North Whitehead, of whom McCulloch was a dedicated reader, recognized schizophrenia as a form of paralogical reasoning highlighting the otherwise logical structure of the mind, McCulloch's choice was in many ways a reflection of how that disorder stood as a condition so opaque that it challenged psychiatry's very capacity to theorize mental illness (Herman 1995; Woods 2011). In 1945, McCulloch himself made a similar argument by notably relating the history of modern psychiatry to that of schizophrenia. In a piece titled "The Modern Concept of Schizophrenia," McCulloch specifically emphasized how the recognition of schizophrenia as a unified disease coincided with the reconceptualization of mental illness as resulting from faulty albeit treatable biological processes (Meduna and McCulloch 1945, 147–49). But more importantly, it is in that piece that McCulloch first framed his model as capable of reducing schizophrenia to a biological basis, by notably framing the latter as the ultimate degree of deviation from the idealized model of neural organization embodied by neural networks. By insisting that such deviations could only be resolved through an active reorganization of neural connections, it was thus as a clinical rationale that McCulloch first deployed his model—one supporting the use of intrusive treatments intervening directly into the brain of psychiatric patients, including electroshock therapy.

Despite being used in half of psychiatric institutions in the United States by the early 1940s, electroshock therapy was still devoid of any strong theoretical backing (Sadowsky 2016; Shorter and Healy 2007). Seeing a clear correspondence between the subjection of the brain to electrical impulses and his reconceptualization of brain activities in terms of neural signals, McCulloch contacted the Josiah Macy Jr. Foundation—then a leading funding body for medical research—and proposed to carry out a clinical trial dedicated to establishing the causes of electroshock therapy's alleged effectiveness. Enticed by his promise to reduce mental conditions such as schizophrenia to "the types of mechanisms" embodied by neural networks (McCulloch 1942, 2), the Macy Foundation promptly approved McCulloch's request and awarded him in 1944 the necessary funding to lead a multiyear, large-scale clinical trial on the use of electroshock therapy to treat patients with schizophrenia. Carried out at the INI, the trial involved subjecting a total of twenty-five patients diagnosed with that condition to electroshocks three times a week over a period of several years. In default of measuring electrophysiological variations directly into the brain, McCulloch relied on blood and urine samples from these patients to identify metabolic markers for both short- and long-term neural alterations.

After one year of such treatments, McCulloch reported that electroshock therapy has had highly uneven effects on the trial's participants, with some of them having seen their condition improve while others not so much. Based on that, McCulloch theorized that what was called schizophrenia was in fact two distinct diseases, each the product of a distinct imbalance in the brain's physiochemical composition. On the one hand, McCulloch speculated that patients who were "'cured' by electric shock" (1948, 3) were victims of a faulty neural system, which led them to inadequately process outside information. On the other hand, McCulloch hypothesized that those who were unresponsive to it were instead affected by a faulty metabolism impinging on the proper transmission of neural signals and recommended the use of other treatments like insulin therapy or $CO_2$ poisoning to alter their brain's chemical composition. While ultimately disproved, this distinction highlighted how McCulloch's assumptions vis-à-vis the physical basis of mental illness gave rise to a model of medical diagnosis overly centered on the functional and physical organization of neural activities, to the point of altogether defining mental conditions based on their receptivity to some somatic treatments over others. Not only that but, by framing these diseases as deviations from the idealized model of neural organization his neural networks embodied, McCulloch simultaneously deployed his model as a rationale for highly intrusive and violent forms of treatment that directly targeted patients' brains.

As such, by embodying an idealized model of neural organization "whose aberrations are our most vexing problems" (McCulloch 1945, 2), neural networks provided a powerful clinical rationale to expand the reach of somatic treatments in the context of wartime psychiatry. In their initial articulation, and before they would be taken up by McCulloch's more computationally inclined colleagues, neural networks conveyed the fundamental assumptions that all mental disorders were essentially physical illnesses treatable through physical means. Therefore, even in the case of diseases that were yet to be provided with a satisfying explanation, neural networks rendered the bodies and more specifically brains of psychiatric patients available for a wide range of experimental manipulations, whose success was measured not so much by their capacity to conclusively treat patients' conditions as by how they provided psychiatrists with new tools to both classify and manage these diseases. For neurons to be constructed as relays and mental states to be reduced to "the passage of alternating currents through the brain" (McCulloch and Pfeiffer 1949, 369–70), neural networks thus needed to not only construct certain subjects as deviating from the organizational ideals they embodied but also literally expose them to "the passage of alternating currents." As such, not unlike Ramón y Cajal's deployment of his neural networks to restore a certain degree of difference between Europeans and the inhabitants of their colonies, McCulloch's own version of this model stood as a highly performative model of the mind whose universalism was proclaimed by experimentally intervening on brains that supposedly deviated from the principles neural networks embodied.

## Managing Intelligence: On Networks and Organization in Late Twentieth-Century Computer Science

Neural networks' third and current formulation as a highly operational, statistical approach to artificial intelligence (AI) is the one that comes to mind to most contemporary commentators. As a computationally intensive model reliant on expensive GPUs, large-scale information-processing infrastructures, and extensive databases, neural networks are generally discussed in terms of the natural resources they consume (e.g., Hogan 2015), the human labor they involve (e.g., Atanasoski and Vora 2019), and the reconfiguration of all things into data they foster (e.g., Ricaurte 2019). Taken together, these different interventions all point to how neural networks' current iteration functions as something other than the data-processing, number-crunching technology it has been made to be. To operate, neural networks rely on the careful management of huge amounts of resources, people, and data on a highly distributed albeit purposefully managed model, which itself mirrors their distributed architecture and purposeful training. Far from peripheral, this managerial component reflects not only how neural networks function as an organizational technology but also how their rearticulation into an AI model was precisely driven by renewed concerns with organization. For contrarily to its previous forms, neural networks' third iteration wasn't so much deployed to intervene on locatable, othered bodies as to organize dispersed networks of working bodies into a productive whole. While seemingly suppressed, concerns with human difference were recast into new ideas about human–machine difference, which mediated how the intelligence necessary for the completion of complex tasks was relocated from the workers themselves to the broader structures they occupied. Before they would raise the prospect of full-on automation, it was thus the goals of optimizing limited resources, enhancing workers' output, and creating new links that neural networks embodied—all things with great appeal to a highly decentralized state like Canada, where their third iteration emerged.

When Canadian funding bodies and research institutions adopted neural networks as a promising new approach to AI research, they did so while embracing a broader understanding of networks as the ideal structure to carry out that type of work. At the time, research networks figured as a key innovation in terms of both coordinating dispersed sites and steering them toward desired collective outcomes. When first championed by the Canadian Institute for Advanced Research (CIFAR), networks were recognized as a promising approach to support cutting-edge research despite the country's limited resources and sheer size. Faced with limited funding prospects, CIFAR developed in 1982 a decentralized, cross-appointment model to fund its inaugural program, which specifically aimed to build capacity in the field of AI throughout the country. When appointed, CIFAR's fellows remained at their host institution but were exempted from their usual university duties and were required to collaborate with one another on CIFAR's nationwide AI program. While born out of necessity, this decentralized, network-like model represented for many public institutions a promising solution to the country's growing innovation lag. Singling out CIFAR's AI program as the paradigmatic example of this new model, the 1982–1984 Royal Commission on the Economic Union and Development Prospects not only described such research networks as a promising approach to move Canada "to the forefront of technological innovation" (Macdonald 1985b, 205) but also emphasized AI's potential contribution to fostering such breakthroughs. Described as a technology uniquely positioned to "substitute [. . . for] human mental efforts" (Macdonald 1985a, 117), AI represented for the commissioners an unprecedented opportunity to free Canadians from laborious office jobs in favor of more creative work.

In the wake of the commission, CIFAR became a regular beneficiary of the Canadian federal government, which recognized its key role in popularizing the idea that research is best carried out when organized into "networks of people [. . .] webbed together by telecommunications and modern transportation" (1987 report quoted in Henderson et al. 2004, 6). It is thanks to that new, steady source of funding that CIFAR not only established new research strands but also expanded its AI program to include scholars similarly concerned with the optimal organization of complex networks. Of the many international scholars provided with a university appointment in Canada by CIFAR, Geoffrey Hinton figured among the handful of fellows who worked on neurologically inspired approaches to AI. After graduating in the late 1970s with a dissertation on how "relational networks containing [. . .] nodes and relations of various types" can constitute "a kind of grammatical knowledge" (1977, 13), Hinton redefined neural networks as a structurally inclined alternative to so-called symbolic approaches to AI. While the latter advanced that cognition necessarily involved operations on symbols (e.g., Fodor 1980; Newell 1980; Pylyshyn 1984), Hinton for his part posited that the currency of cognition was instead "neuron-like computing elements" organized into networks (Hinton, McClelland, and Rumelhart 1986, 77), in which it was the "interacting influences among the units [. . . that was] doing the work" (1986, 85).

From Hinton's point of view, the objective of AI was thus not so unlike that of CIFAR itself. In both cases, the challenge consisted in finding the best method to organize limited resources—be they computational ones, in Hinton's case, or human ones, in CIFAR's—as "to get the most possible out of a simple network of connected units" (1986, 87). Hinton's appointment at CIFAR and move to the University of Toronto in 1986 even coincided with his introduction of precisely such a method—one designed to optimize the connections among units so the task at hand was "encoded by a pattern of activity in many units rather than by a single active unit" (1986, 88). Known as "backpropagation," this procedure consisted in "repeatedly adjust[ing] the weights of the connections in the network" as to minimize the difference between a net's actual and desired output for a given input (Rumelhart, Hinton, and Williams 1986, 533). While dismissed by Hinton's CIFAR peers as bearing little resemblance to how the brain functions let alone thinks (see Fodor and Pylyshyn 1988; Levesque 1989), this method nevertheless provided a highly functional approach to embedding complex operations into simple networks. If this method's first use case—namely, "learning the associations of 20 pairs of random binary vectors of length 10" (Plaut, Nowlan, and Hinton 1986, 20)—bore little practical application, it still provided an example of how a complex task could be performed at the level of the network itself, without being reducible to any of its individual components.

But beyond such individual applications, it is the very prospect of restoring a certain degree of control within similarly distributed systems that appealed to CIFAR's public sponsors. In a landmark report published the year following Hinton's appointment, the Government of Canada singled out AI as a technology uniquely positioned to improve government operations by providing "substitutes for [human] expertise" (McKinnon 1987, 4). By raising the prospect of decreasing the government's reliance on skilled work, AI was framed as a promising way to relocate the intelligence and knowledge necessary for the completion of complex government operations at the level of institutions themselves, at the expense of the workers who would no longer hold the necessary knowledge to bear judgment or act on them. In the image of neural networks themselves, what emerged was a view of government operations as independent from any of its individual components, with both the deskilling and reorganization of low-ranking civil servants being recast as a privileged way to steer an organization as large as the Canadian civil service toward desired ends.

The type of work pursued by Hinton in the following years with the support of both provincial and federal bodies directly mirrored this broader ambition of not so much replacing human workers as breaking down their tasks into smaller, more standardized ones. With grants from the Information Technology Research Center to research speech recognition, Hinton developed a time-delay neural network architecture designed to identify phonemes in sound samples of different lengths (Lang, Waibel, and Hinton 1990, 42), which could be used to create phonetic transcriptions that could be then polished by human workers. With another grant from the same body, Hinton trained a neural network to recognize handwritten numbers on envelopes (Hinton, Williams, and Revow 1992, 54–56), with potential applications in the classification and distribution of mail. With the support of the Canadian Institute for Robotics and Intelligent Systems, Hinton developed a neural network that could turn hand gestures into computer-readable inputs (Fels and Hinton 1998), which could be used to both build new adaptive interfaces and monitor the actions of those using them. While none of these applications were overtly political, they nevertheless raised the prospect of doing away with various forms of contextual or local knowledge by reducing the contribution of human workers to that of either polishing or executing computer outputs. In the name of "increase[ing] productivity in government operations" (McKinnon 1987, 13), these applications provided new ways to streamline operations, centralize decision making, and discipline workers. If neural networks provided both the necessary tools and metaphors to reconceive of a large and dispersed workforce as itself a network of simple units, it is then because their development was part of a broader attempt to deskill and deprofessionalize workers, making it easier to train, manage, and if need be, replace them.

Back at CIFAR, it is this very capacity of Hinton and his operational approach to build connections with public funders that led to a similar streamlining of the institute's AI program. As the latter's director then noted, CIFAR's efforts to "facilitate the transfer of knowledge and the development of a strong applied research sector" (Zenon Pylyshyn quoted in Brown 2007, 64) had proven mostly unsuccessful by the turn of the 1990s, with Hinton's work constituting a much-welcomed exception to this shortcoming. In the subsequent years, the capacity of fellows to secure links and partnerships with both private and public bodies gradually became the main source of validation for their work, greatly favoring those pursuing more applied lines of research. While many of CIFAR's original fellows saw their affiliation revoked by the mid-1990s, the new fellows were, for the most part, selected from among the new generation of computer scientists working on neural networks and their application in subfields like computer vision and speech recognition. Despite the multidisciplinary ambitions of CIFAR's network-like model, it was precisely this emphasis on building connections as an end in itself that ultimately led to a growing homogenization around the nodes that proved the most successful at it. In the image of neural networks, CIFAR's network approach did not then provide a model for the creation of new links; instead, it advanced a new vision of interconnectedness and interdependence as key mechanisms to create cohesion out of difference. At the same time as it promoted collaboration and transdisciplinarity, CIFAR thus also provided a rich example of how clear research goals can be embedded into the very organization of research networks.

By providing a highly operational approach to machine intelligence as well as embodying the idealized structure in which to carry that type of work, neural networks' third iteration then emerged as a solution to the problem of how best to organize distributed networks so control and cohesion can be restored. In this specific articulation, and before they would be relayed to a purely computational framework, neural networks promoted the idea that a certain degree of oversight could be restored in distributed systems by simplifying the contribution of its individual components, promoting connections as an end in itself and relocating the system's functions at the level of its structure. That the human workers involved in formatting the 768 sound samples used for Hinton's speech recognition model or the 60,000 images used for his later work on computer vision received but a passing remark (see Lang, Waibel, and Hinton 1990, 25–26; Krizhevsky 2009, 32) might come as no surprise; and yet, it is symptomatic of the broader devaluation of local, contextual, and situated knowledge fueled by this model's reconceptualization of all its components as "simple units." For Hinton to recast AI's aim as that of identifying how best to organize "internal units [. . . so they] come to represent important features of a task" (Hinton 1990, 185), it was then necessary for all the actors, resources, and organizations involved in neural networks' rearticulation, development, and implementation to be recast in similar structural terms. More than a biologically inspired model of machine intelligence, neural networks thus emerged in its current form as a structurally inclined approach to organization, in which control was exercised by not only rendering complex social contexts into networks of simple units but also acting upon them as such.

## The Making of Neurality

When reviewed alongside one another, the various practices and discourses surrounding neural networks' three main iterations in the context of nineteenth-century histology, wartime psychiatry, and late twentieth-century computer science undermine the teleological narrative that is now associated with this model. Far from showing the gradual yet steady convergence of human and machine intelligence, these three moments highlight how neural networks were deployed to not only answer widely different questions but also manage, more than explain, the processes and functions they theorized. If anything, what bridges these three moments is a sustained investment in experimentation manifested by the development of new standardized techniques to intervene on individual neurons, whole brains, and expansive networks of people. At the same time as they promoted the idea that "it seems best to handle even apparent continuities as some numbers of some little steps" (McCulloch and Pfeiffer 1949, 368), all these techniques were deployed to manipulate, and experiment on, these fundamental structures until they produced the desired output—namely, restoring a certain degree of control in the face of colonial dissolution, deinstitutionalization, and political decentralization. Far from being the abstract model of the mind applicable to both humans and machines they have been made to be, neural networks instead stand as a complex experimental construct produced through the sustained manipulation, control, and management of both the objects and subjects deviating from the principles they embody.

What the history of neural networks points to therefore goes beyond the question of whether representing biological neurons as simulated units probabilistically linked together is accurate or not, or even if neurology indeed provides a useful lens to understand intelligence. Instead, what it shows is the recurrence, and repeated failure, of a certain way of reducing complex phenomena to locatable structures available for manipulation. Materialized through practices as varied as silver staining and colonial health policies, brain lacerations and electroshock therapy, and organizational reforms and the programming of software-simulated neurons, neural networks captured things like neurons, brains, and whole social systems as substrates for the circulation of new structuralist ideas about organization, communication, and management. In doing so, far from providing a final explanation for any of these entities, neural networks introduced just enough indeterminacy to allow for their own inscription onto various brains and bodies, to the point that they no longer appeared as concepts but instead as empirical facts. In that sense, for the way they create both the channel and the receiving end for the circulation of the ideas they embody, I propose to understand neural networks not only as a medium but also as one that is representative of a broader class I term *neural media,* which are defined by the way they communicate a technologized view of the human biology, psychology, and sociality.

From Ramón y Cajal's racialized brain and McCulloch's diseased mind to Hinton's networked intelligence, neural networks have systematically mediated how various bodies were perceived, known, and made use of. Once constructed as the site of human difference, the brain was used to justify the subjection of various populations to new forms of colonial management; once recognized as the site of mental deviance, the mind allowed psychiatrists to capture certain bodies as receptive to highly intrusive clinical treatments; once recast as the outcome of a certain way of organizing limited computational resources, intelligence became a privileged lens to devalue the situated knowledge and expertise of low-ranking workers. In all these cases, brains, minds, and intelligence were not only made to convey certain meanings and enable certain practices by being mobilized within the neural domain but also allowed for that domain's primacy to be affirmed across a variety of fields, time periods, and national settings. In that context, what neural media theorizes is how neurality—here understood as the idea that all mental states can be reduced to locatable neural structures—is itself a highly performative construct, which promotes organizational principles such as atomization, interdependence, and mediation at the same time as it frames them as biologically given.

This reading builds on various accounts of how conceptions of machine intelligence have been closely associated with the management of bodies and workers—including Ranjodh Singh Dhaliwal's recent discussion of computation's historical involvement in the racialized organization of class and labor (2022; see also Nakamura 2014; Rhee 2018)—but more specifically highlights how neurality itself has come to stand for a broader organizational ethos. That atomization, interdependence, and mediation are now so closely associated with the structure of the brain is but the outcome of sustained efforts to project onto human biology principles, processes, and structures inherited from race science, wartime psychiatry, and new public management. Once reduced to its structure, the brain became a powerful device to naturalize a vision of organization as the main determinants of systems and thus the privileged means through which social orders can be (re)produced. If simple units and interconnected networks suddenly proliferate when neural networks are involved, it is not so much because of their universality as structures but instead because of the performative nature of neurality itself, which mediates how complex biological, psychological, and social systems are made legible as, and actively shaped into, dense networks of tightly interconnected units.

When considered in relation to neural networks' trajectory across histology, psychiatry, and computer science, what neural media show is how the history of artificial intelligence is essentially that of mediating who or what is recognized as the bearer of intelligence. For in each of neural networks' iterations, intelligence has been systematically constructed as a faculty unevenly exhibited. When reduced to a purely structural basis, intelligence becomes a quality inherent to those who embody the right structure at the expense of those deviating from it. The type of intelligence posited by such structuralist views is one completely emptied out of its content, context, and conditions, but which simultaneously mirrors in its form and functions the stratified worldview that underpins its attribution to some versus others. That AI participates so actively in the reproduction of historical divides around ideas about racial, pathological, and human–machine difference is but a reflection of how intelligence has been constructed as held by certain people more than others, with all its iterations both biological and computational being at the center of broader structures of power that maintain that uneven distribution of intellectual potential. What neural media point to is, then, how the very attribution of intelligence is actively mediated by various technologies, which naturalize certain bodies as its privileged bearers while obfuscating the politics behind that attribution. More than a century after neurons were first constructed as intelligence's fundamental mediator, it is now the architecture of biologically inspired learning machines that fulfills the function of communicating such a technologized view of human biology, psychology, and sociality, pointing to the persistence of neurality itself as both an artifact of, and an artifice for, the attribution of intelligence across humans and machines.

# Notes

This chapter is part of a broader project I have been pursuing for the past few years. As such, it has been shaped and informed by the invaluable insights of many people, including Wendy Hui Kyong Chun, Lukas Rieppel, and Gertrud Koch. In writing and revising this chapter, I also benefitted from the intellectual generosity and, I'm afraid, bottomless patience of my coauthors, Ranjodh Singh Dhaliwal and Lucy Suchman, who read versions of it at times when they barely qualified as drafts. Their comments and interventions have shaped my thoughts in a way that will influence my work way beyond the scope of this chapter. Finally, I am very much grateful to the two anonymous reviewers who read our book first for their intellectual generosity and wish to convey my most sincere gratitude to Mercedes Bunz, who oversaw the completion of this book.

1 All quotes from texts in French and Spanish are translations by the author.

2 If machine learning stands more generally for a statistical approach to AI centered on the extraction of patterns from large amounts of data, then neural networks can be defined as a biologically inspired model that relies on probabilistically weighted neuron-like units to identify such patterns.

3 For many, Siskin's definition will seem a bit unusual. Canonical definitions of mediation generally characterize the latter in terms of "social agencies [. . .] deliberately interpos[ing] between reality and social consciousness" (Williams 1976, 206) or of the appropriation of "the techniques, forms, and social significance of other media [. . .] to rival or refashion them in the name of the real" (Bolter and Grusin 1999, 65). What Siskin's definition does, in comparison, is to emphasize both the material aspect of that process and how it physically intervenes on the settings where it unfolds.

4 I am grateful to Paolo Mazzarello, the leading scholar on Golgi, for having confirmed the origins of the human tissues used by Golgi for his histological studies. As he explained to me in a private communication, all the tissues Golgi worked with were from the Hospital of San Matteo and the Hospital for the Chronically Sick at Abbiategrasso, the two main medical institutions where he worked during his career.

5 It is worth noting that the term "Creole" was undergoing a profound transformation in the second half of the nineteenth century. While initially referring to people of Spanish descent born in colonial America, "Creole" had come to encompass all America-born people independently of their ethnic background by the time the first Cuban war of independence broke out (see, for instance, B. Anderson 2006; Simon 2017). This change was symptomatic of growing tensions between the colonies and peninsular Spain, which had prohibited all Creoles—including those of Spanish descent—to occupy any positions in Cuba's colonial government.

# References



Abraham, Tara. 2016. *Rebel Genius: Warren S. McCulloch's Transdisciplinary Life in Science.* Cambridge, Mass: MIT Press.

Anderson, Benedict. 2006. "Creole Pioneers." In *Imagined Communities: Reflections on the Origin and Spread of Nationalism,* 2nd ed., 47–66. London: Verso.

Anderson, Warwick. 2006. *Colonial Pathologies: American Tropical Medicine, Race, and Hygiene in the Philippines.* Durham, N.C.: Duke University Press.

Atanasoski, Neda, and Kalindi Vora. 2019. *Surrogate Humanity: Race, Robots, and the Politics of Technological Futures.* Durham, N.C.: Duke University Press.

Bolter, Jay, and Richard Grusin. 1999. *Remediation: Understanding New Media.* Cambridge, Mass.: MIT Press.

Bonin, Gerhardt von, Hugh Garol, and Warren McCulloch. 1942. "The Functional Organization of the Occipital Lobe." *Biological Symposia* 7: 164–92.

Brown, Craig. 2007. *A Generation of Excellence: A History of the Canadian Institute for Advanced Research.* Toronto: University of Toronto Press.

Chia, Aleena. 2022. "Oneirogenic Innovation in Consciousness Hacking." In *Techno-pharmacology* by Joshua Neves, Aleena Chia, Susanna Paasonen, and Ravi Sundaram, 55–88. Minneapolis: meson press and University of Minnesota Press.

Dhaliwal, Ranjodh. 2022. "The Cyber-Homunculus: On Race and Labor in Plans for Computation." *Configurations* 30, no. 4: 377–409.

Dupuy, Jean-Pierre. 1994. *Aux origines des sciences cognitives.* Paris: La Découverte.

Dusser de Barenne, Johannes, Hugh Garol, and Warren McCulloch. 1941. "The 'Motor' Cortex of the Chimpanzee." *Journal of Neurophysiology* 4, no. 4: 287–303.

Dusser de Barenne, Johannes, and Warren McCulloch. 1939. "Physiological Delimitation of Neurons in the Central Nervous System." *American Journal of Physiology* 127: 620–28.

Fels, Sidney, and Geoffrey Hinton. 1998. "Glove-TalkII—A Neural-Network Interface which Maps Gestures to Parallel Formant Speech Synthesizer Controls." *IEEE Transactions on Neural Networks* 9, no. 1: 205–212.

Fitch, Frederic. 1944. "Review of 'A Logical Calculus of the Ideas Immanent in Nervous Activity' by Warren S. McCulloch and Walter Pitts." *The Journal of Symbolic Logic* 9, no. 2: 49–50.

Fodor, Jerry. 1980. *The Language of Thought.* Cambridge, Mass.: Harvard University Press.

Fodor, Jerry, and Zenon Pylyshyn. 1988. "Connectionism and Cognitive Architecture: A Critical Analysis." *Cognition* 28, no. 1–2: 3–71.

Gerlach, Joseph von. 1872. "Über Die Struktur Der Grauen Substanz Des Menschlichen Grosshirns." *Zentralbl. Med. Wiss* 10: 273–75.

Golgi, Camillo. 1869a. "Sull'eziologia delle alienazioni mentali in rapporto alla prognosi ed alla cura." *Annali universali di medicina* 207: 564–632.

Golgi, Camillo. 1869b. "Sulla struttura e sullo sviluppo degli psammomi." *Morgagni* 11: 874–86.

Golgi, Camillo. 1873. *Sulla struttura della sostanza gricia del cervello.* Milan, Italy: Stabilumento dei fratelli Rechiedei.

Golgi, Camillo. 1874. "Sulle alterazioni degli organi centrali nervosi in un caso di corea gesticolatoria associata ad alienazione mentale." *Rivista clinica* 2, no. 4: 361–77.

Golgi, Camillo. 1907. *La doctrine du neurone : Théorie et faits.* Stockholm, Sweden: Imprimerie royale P.-A. Norstedt & Fils.

Grob, Gerald. 1983. *Mental Illness and American Society, 1875–1940.* Princeton, N.J.: Princeton University Press.

Henderson, Mark, Monica Hughes, Debbie Lawes, Kathie Lynas, and Joe Sornberger. 2004. "Excellence Has No Fixed Address: The NCE Pioneers." Ottawa, Canada: Networks of Centres of Excellence Canada.

Herman, Ellen. 1995. *The Romance of American Psychology: Political Culture in the Age of Experts.* Berkeley: University of California Press.

Hinton, Geoffrey. 1977. "Relaxation and Its Role in Vision." PhD diss. University of Edinburgh.

Hinton, Geoffrey. 1990. "Connectionist Learning Procedures." *Artificial Intelligence* 40: 185–234.

Hinton, Geoffrey, James McClelland, and David Rumelhart. 1986. "Distributed Representations." In *Parallel Distributed Processing: Explorations in the Microstructure of Cognition, Volume 1,* ed. David Rumelhart and James McClelland, 77–109. Cambridge, Mass.: MIT Press.

Hinton, Geoffrey, Christopher Williams, and Michael Revow. 1992. "Combining Two Methods of Recognizing Hand-Printed Digits." *Artificial Neural Networks* 2: 53–60.

Hogan, Mél. 2015. "Data Flows and Water Woes." *Big Data & Society* 2, no. 2: 1–12.

Jagoda, Patrick. 2020. *Experimental Games: Critique, Play, and Design in the Age of Gamification.* Chicago: University of Chicago Press.

Kay, Lily. 2001. "From Logical Neurons to Poetic Embodiments of Mind: Warren S. McCulloch's Project in Neuroscience." *Science in Context* 14, no. 4: 591–614.

Kramer, Paul. 2006. "Blood Compacts: Spanish Colonialism and the Invention of the Filipino." In *The Blood of Government: Race, Empire, the United States, and the Philippines.* Chapel Hill: The University of North Carolina Press.

Krizhevsky, Alex. 2009. "Learning Multiple Layers of Features from Tiny Images." Toronto, Canada. https://www.cs.toronto.edu/~kriz/learning-features-2009-TR.pdf.

Lang, Kevin, Alex Waibel, and Geoffrey Hinton. 1990. "A Time-Delay Neural Network Architecture for Isolated Word Recognition." *Neural Networks* 3: 23–43.

Levesque, Hector. 1989. "Making Believers out of Computers." In *Readings in Artificial Intelligence and Databases,* ed. John Mylopolous and Michael Brodie, 69–82. San Mateo, Calif.: Morgan Kaufmann Publishers.

Macdonald, Donald. 1985a. *Report of the Royal Commission on the Economic Union and Development Prospects for Canada: Volume 1.* Ottawa, Canada: Privy Council Office.

Macdonald, Donald. 1985b. *Report of the Royal Commission on the Economic Union and Development Prospects for Canada: Volume 2.* Ottawa, Canada: Privy Council Office.

Mauss, Marcel. 1923. "Essai sur le don : Formes et raisons de l'échange dans les sociétés archaïques." *L'Année sociologique* 1: 30–180.

Mazzarello, Paolo. 2009. *Golgi: A Biography of the Founder of Modern Neuroscience.* Oxford: Oxford University Press.

McCulloch, Warren. 1941. "Letter to Percival Bailey from February 4, 1941." In

Warren S McCulloch Papers, Series No. 35/3/99, Box 2. Champaign: University of Illinois Archives.

McCulloch, Warren. 1942. "Letter to Frank Fremont-Smith from June 24, 1942." In Warren S McCulloch Papers, Series No. 35/3/99, Box 8, 1–3. Champaign: University of Illinois Archives.

McCulloch, Warren. 1945. "Letter of Recommendation in Support of Walter Pitts' Application to the Guggenheim Memorial Foundation (1945?)." In Warren S McCulloch Papers, Series No. 35/3/99, Box 19, 1–2. Champaign: University of Illinois Archives.

McCulloch, Warren. 1948. "Letter to Frank Fremont-Smith from June 1, 1948." In Warren S McCulloch Papers, Series No. 35/3/99, Box 8, 1–9. Champaign: University of Illinois Archives.

McCulloch, Warren. 1974. "Recollections of the Many Sources of Cybernetics." *ASC Forum* 5, no. 16: 1–26.

McCulloch, Warren, and John Pfeiffer. 1949. "Of Digital Computers Called Brains." *The Scientific Monthly* 69, no. 6: 368–76.

McCulloch, Warren, and Walter Pitts. 1943. "A Logical Calculus of the Ideas Immanent in Nervous Activity." *Bulletin of Mathematical Biophysics* 5: 115–33.

McKinnon, Peter. 1987. *An Overview of AI & Recent Related Government Activities.* Ottawa, Canada: Ministry of State for Science and Technology of Canada.

Meduna, Ladislas, and Warren McCulloch. 1945. "The Modern Concept of Schizophrenia." *Medical Clinics of North America* 29, no. 1: 147–64.

Meynert, Theodor. 1872. "Vom Gehirne Der Säugetiere." In *Handbuch Der Lehre von Den Geweben Des Menschen Und Thiere,* ed. Salomon Stricker, 694–808. Leipzig, Germany: Engelmann.

Milburn, Colin. 2010. "Modifiable Futures Science Fiction at the Bench." *Isis* 101: 560–69.

Nakamura, Lisa. 2014. "Indigenous Circuits: Navajo Women and the Racialization of Early Electronic Manufacture." *American Quarterly* 66, no. 4: 919–41.

Neill, Deborah. 2012. *Networks in Tropical Medicine: Internationalism, Colonialism, and the Rise of a Medical Specialty, 1890–1930.* Stanford, Calif.: Stanford University Press.

Neumann, John von. 1951. "The General and Logical Theory of Automata." In *Cerebral Mechanisms in Behavior: The Hixon Symposium,* ed. Lloyed Jeffress, 1–41. New York: John Wiley and Sons.

Neves, Joshua. 2022. "The Internet of Things and People." In *Technopharmacology* by Joshua Neves, Aleena Chia, Susanna Paasonen, and Ravi Sundaram, 89–119. Minneapolis: meson press and University of Minnesota Press.

Newell, Allen. 1980. "Physical Symbol Systems." *Cognitive Science* 4: 138–83.

Palay, Sanford, and George Palade. 1955. "The Fine Structure of Neurons." *Journal of Biophysical and Biochemical Cytology* 1, no. 1: 69–107.

Pittaluga, Gustavo. 1910. "La glossina palpalis en la colonia española." In *Informe de la comisión del Instituto nacional de higiene de Alfonso XIII enviada a las posesiones españolas del Golfo de Guinea para el estudio de la enfermedad del sueño y de las condiciones sanitarias de la colonia,* ed. Gustavo Pittaluga, 393–402. Madrid, Spain: Blass.

Pitts, Walter. 1942a. "Some Observations on the Simple Neuron Circuit." *Bulletin of Mathematical Biophysics* 4, no. 3: 121–29.

Pitts, Walter. 1942b. "The Linear Theory of Neuron Networks: The Static Problem." *Bulletin of Mathematical Biophysics* 4, no. 4: 169–75.

Plaut, David, Steven Nowlan, and Geoffrey Hinton. 1986. "Experiments on Learning by Back Propagation." In *Technical Report CMU-CS-86-126.* Pittsburgh: Carnegie-Mellon University.

Pylyshyn, Zenon. 1984. *Computation and Cognition: Toward a Foundation for Cognitive Science.* Cambridge, Mass.: MIT Press.

Ramón y Cajal, Santiago. 1887. "Notas de laboratorio: Tejido óseo y coloración de los cortes de hueso." *Boletín médico valenciano.* 20: 1-8.

Ramón y Cajal, Santiago. 1888a. "Estructura de los centros nerviosos de las aves." *Revista trimestral de histología normal y patológica* 1: 1–10.

Ramón y Cajal, Santiago. 1888b. "Sobre las fibras nerviosas de la capa molecular del cerebelo." *Revista trimestral de histología normal y patológica* 2: 1–17.

Ramón y Cajal, Santiago. 1894. "The Croonian Lecture: La fine structure des centres nerveux." *Proceedings of the Royal Society* 55: 444–68.

Ramón y Cajal, Santiago. 1897. *Manual de histología normal y técnica micrográfica para uso de estudiantes.* 2nd ed. Madrid, Spain: Imprenta y librería Nicolás Moya.

Ramón y Cajal, Santiago. 1904. *Elementos de histologia normal y de técnica micrográfica.* Madrid, Spain: Moya.

Ramón y Cajal, Santiago. 1907. "Structure et connexions des neurones." *Nordiskt Medicinskt Arkiv* 1, no. 2: 1–28.

Ramón y Cajal, Santiago. 1910. "Prólogo." In *Informe de la comisión del Instituto nacional de higiene de Alfonso XIII enviada a las posesiones españolas del Golfo de Guinea para el estudio de la enfermedad del sueño y de las condiciones sanitarias de la colonia,* ed. Gustavo Pittaluga, 1–12. Madrid, Spain: Blass.

Ramón y Cajal, Santiago. 1989 [1901]. *Recollections of My Life.* Trans. E. Horne Craigie. Cambridge, Mass.: MIT Press.

Ramón y Cajal, Santiago. 1999 [1897]. *Advice for a Young Investigator,* 4th ed. Trans. Neely Swanson. Cambridge, Mass.: MIT Press.

Rhee, Jennifer. 2018. *The Robotic Imaginary: The Human and the Price of Dehumanized Labor.* Minneapolis: University of Minnesota Press.

Rheinberger, Hans-Jörg. 1997. *Toward a History of Epistemic Things: Synthesizing Proteins in the Test Tube.* Stanford, Calif.: Stanford University Press.

Ricaurte, Paola. 2019. "Data Epistemologies, The Coloniality of Power, and Resistance." *Television & New Media* 20, no. 4: 350–65.

Rumelhart, David, Geoffrey Hinton, and Ronald Williams. 1986. "Learning Representations by Back-Propagating Errors." *Nature* 323: 533–36.

Rumelhart, David, and James McClelland. 1987. *Parallel Distributed Processing: Explorations in the Microstructure of Cognition.* Cambridge, Mass.: MIT Press.

Rumelhart, David, James McClelland, and Geoffrey Hinton. 1986. "The Appeal of Parallel Distributed Processing." In *Parallel Distributed Processing, Volume 1,* ed. David Rumelhart and James McClelland, 3–44. Cambridge, Mass.: MIT Press.

Rumelhart, David, and David Zipser. 1986. "Feature Discovery by Competitive Learning." In *Parallel Distributed Processing, Volume 1,* ed. David Rumelhart and James McClelland, 151-193. Cambridge, Mass.: MIT Press.

Sadowsky, Jonathan. 2016. *Electroconvulsive Therapy in America: The Anatomy of a Medical Controversy.* London: Routledge.

Shorter, Edward, and David Healy. 2007. *Shock Therapy: A History of Electroconvulsive Treatment in Mental Illness.* New Brunswick, N.J.: Rutgers University Press.

Simon, Joshua. 2017. *The Ideology of Creole Revolution: Imperialism and Independence in American and Latin American Political Thought.* Cambridge: Cambridge University Press.

Siskin, Clifford. 2016. *System: The Shaping of Modern Knowledge.* Cambridge, Mass.: MIT Press.

Spivak, Gayatri. 1999. *A Critique of Postcolonial Reason: Toward a History of the Vanishing Present.* Cambridge, Mass.: Harvard University Press.

Stäheli, Urs. 2020. "Undoing Networks." In *Undoing Networks* by Tero Karppi, Urs Stäheli, Clara Wieghorst, and Lea Zierott, 1–30. Minneapolis: meson press and University of Minnesota Press.

Star, Susan Leigh. 1989. *Regions of the Brain: Brain Research and the Quest for Scientific Certainty.* Stanford, Calif.: Stanford University Press.

Suchman, Lucy. 1987. *Plans and Situated Actions: The Problem of Human–Machine Interaction.* Cambridge: Cambridge University Press.

Weinstein, Deborah. 2013. *The Pathological Family: Postwar America and the Rise of Family Therapy.* Ithaca, N.Y.: Cornell University Press.

Wiener, Norbert. 1945. "Letter to Arturo Rosenblueth from January 24, 1945." In *Norbert Wiener Papers, Box 4, Folder 67,* 1–2. Cambridge, Mass.: MIT ArchivesSpace.

Williams, Raymond. 1976. *Keywords: A Vocabulary of Culture and Society.* New York, N.Y.: Oxford University Press.

Woods, Angela. 2011. *The Sublime Object of Psychiatry: Schizophrenia in Clinical and Cultural Theory.* Oxford: Oxford University Press.

Wynter, Sylvia. 1995. "1492: A New World View." In *Race, Discourse, and the Origin of the Americas: A New World View,* ed. Vera Lawrence Hyatt and Rex Nettlefort, 5–57. Washington, D.C.: Smithsonian Institution Press.

[2]

# On Parascientific Mediations

## Science Fictions, Educational Platforms, and Other Substrates That Think Neural Networks

**Ranjodh Singh Dhaliwal**

What happens when technoscience is off duty? The scientific bench and the computational Integrated Development Environment (IDE) are each, at the end of the day, a platform where objects, ideas, and people come and go. The oft-expected procedurality of science and technology—inscription systems that produce research papers, the peer review process, the speculations and hopes concocted in grant proposals, even the social networks engendered by the scientists and technologists—is always surrounded by so much else. Conditions of growth and sustenance—literally media in the biological sense if you are growing living matter in the laboratory—exert various pressures and perform various roles of their own. This chapter is about one such genre of media: parascientific media. Parascientific media (extending Kaplan and Radin 2011), for me, are forms of media that surround what is considered the proper scientific apparatus. Such media, I argue in this chapter, are used by scientists and technologists to think through their technoscientific problems and to convince others of their scientific positions. In order to show what parascientific media are and how they operate, the chapter looks at two distinct nodes in the history of neural networks. Working through, around, and more importantly after two controversial moments—the neural debates in European pathology around the turn of the twentieth century and the connectionist-symbolist artificial intelligence tug-of-war in computer science and engineering in Silicon Valley around the 2010s—this chapter demonstrates how some oft-unremarked genres of media around science and technology feature in the relationship between the doers (the scientists, coders, granting agencies, experts) and the publics that they interface with. Neural networks here can be understood as kinds of figurations that seek, or perhaps more accurately find themselves in, other substrates where thinking about thinking can happen. Controversies about science and technology may have complicated routings and battles, but parascientific discourse comprises one clear location for their settlement.

Much as in the rest of this book, two distinct histories of two related neural figures collide in the forthcoming pages. On the one hand is the neuron, the OG of the central nervous system, colloquially understood as the thinking cell in most animal species. On the other hand is the artificial neuron, understood analogically as a biomimicry instantiated in silicon computational hardware. Both figures, in this story, are separated by almost a century; this gap is not a historical given—the ties between a neuron and an artificial neuron get inaugurated right in the middle of the twentieth century, after all—but instead a choice made here to illuminate some specific transhistorical features of, and maneuvers in, thinking about thinking.[1] Some of the maneuvers in the coming pages will have applications more general than neural networks. However, to make claims about neurons (both organic and inorganic) is not just to make claims about thought but also about the nature and media of rhetorical claims themselves. In this regard, neural histories are also histories of claims and counterclaims. In fact, as we shall see, the many competing claims about how we think undergird the long history of animating neurons to speak for, and critically think for, some people and not others.[2] Furthermore, this speaking to (and of) is not, crucially enough, always conscious and considered. The two scenes that are called upon here to speak for this piece showcase a slice of neurality that is not completely scientifically determined but instead argued for through the use of certain mediating techniques to solve, stabilize, or at least purport to think through and re-present, technoscientific controversies of their respective day(s) and age(s). My basic point in this piece is that these mediating techniques, quite simply, mediate, often in a covert and unannounced way.

Studying controversies is a time-tested method for scholars of science and technology; moments of uncertainty in science offer some of the finest testing grounds for everything that is taken as an objective natural given.[3] Controversies (from the Latin *contra* ["against"] and *vertere* ["to turn"]), by their very constitution, are about disagreements and disputes; they mark a temporal moment when something is, at least publicly, up in the air. Controversies are only distinguished in the historical record by their absence or by their settled presence (in other words, by what proves, retroactively, to not be a controversy). It is this that excites historians, philosophers, sociologists, and anthropologists of science: the ability to mark a critical move where something that is, was once not, and perhaps could have been not. It is also here that science studies finds itself in the process of being indistinguishable from (at least a certain form of) media archaeology, where the search for lost technologies, dead scientific paradigms and practices, and "what could have been" is a foundational element of the scholarly method (Huhtamo and Parikka 2011). Controversies, in reaching out beyond scientific confines, often create their own publics for resolution.[4] But when a controversy is settled, there is more work to be done and more publics to be cultivated; it is here, in these postcontroversy publics, that a large chunk of the story in this chapter will be staged.

The two controversies at hand here—the disagreements over cellular composition of the brain at the turn of twentieth century and the disagreements over the optimum kind of artificial intelligence paradigm just after the turn of the twenty-first—illuminate not only the contested nature of neural histories but also the working stages of the truth claims that come with certain wins and losses in those contests. Using the specific examples of media—science fiction stories in the first instance and scientifically educational platform-apparatuses in the more recent case—this piece further develops the idea of parascientific media, a term introduced by Sarah Kaplan and Joanna Radin to name publications that intervene in the travels of technical knowledge. In this story, I extend parascientific media, through engagement with media studies, to other senses and kinds of mediation; through an attention to media that may not even be talking explicitly about the science in question but nevertheless is engaged in thinking about its truth claims and epistemic virtues, I seek to interrogate the thinking that happens around the bench and the IDE (see Kaplan and Radin 2011). Denoting the cognitive work done by and through alternate[5] inscriptional forms that neither consciously announce themselves as an immutable mobile (Latour 1986) nor reorient allies (Latour 1993) in those conventional senses, parascientific mediations, when attended to, help recover the environmental role played by these sandboxes and sites of rhetoric while also noting the formal features that enable certain kinds of thinking over others. In this particular story, parascientific media are used to clarify and build publics around a closing debate, converting a "winning" position into a publicly accepted one. Such procedures also involve usually unseen substrates (media, in this case) being used for thinking through the problems of scientific inquiry and epistemology. Be they in the form of science fiction stories, scientific advice books, or educational courseware, parascientific media are not just about science communication, though sometimes there is some overlap, but also often about using the medial qualities for thinking itself.

In the case of neural networks, parascientific mediations make evident the contingency and constructedness of claims about cognition and intelligence by laying bare the translations required to make scientific materials and semiotics into parascientific forms. If my claim about parascientific mediations as one exploratory ground for scientific epistemology is to be accepted, then neural networks show up as scientific objects that are parascientific per se. As demonstrated by Lepage-Richer in chapter 1 of this volume, and Suchman in chapter 3, neural networks are thoroughly social and situated in both their biological and computer-scientific iterations. It is no wonder, then, that neural networks emerge, for me, as a privileged site of investigation for parascientific media. As we shall see in the coming sections, neurons, networks, and neural networks, even when they are settled matters, need to be resituated, rethought, and formally reinscribed over and over again both in and outside the laboratory.

## Scene One: The Strange Fictional Career of Doctor Bacteria

The backdrop for this first scene is the reticular versus neuron theory debate around the turn of the twentieth century already briefly explored in the Introduction to this volume and in Lepage-Richer's rich narrative. While the whole story is, of course, quite complicated, the simplest version runs through two camps with competing ideas: those of the Italian scientist Camillo Golgi and the Spanish scientist Santiago Ramón y Cajal. A scientific controversy (retrospectively construed as such, of course) ensues, ideas compete, one wins over the other, science goes back to normalcy: the exaggerated version of this narrative genre is well known and formally recognizable since Kuhn's analyses (Kuhn 1970). Golgi—who is credited with inventing the staining technique *la reazione nera* (or "the black reaction"), which uses a silver compound to first stain and then image a certain organic tissue—argued, following others (such as the German Joseph von Gerlach, who had used the term "syncytium") before him, that the nervous system is one continuous network (and the two different variations of interpreting networks, one continuous, one noncontinuous, are also of course at play here) with no clear internal distinctions; this was termed the reticular theory. Ramón y Cajal, on the other hand, had used Golgi's staining technique to visualize several parts of the nervous system and inferred the opposite conclusion: that brains were composed of several small cells that were distinct and discontinuous but still connected with each other; this was the neuron doctrine that methodologically stood atop the work of Golgi, which it disagreed with conceptually. As one might imagine, the positions herein—one that posits a giant contiguous unit and one that posits several smaller units in connection—are fundamentally epistemological as much as they are diagrammatic or visual. And they are, of course, also very much visual.

The central question on which the two camps disagreed involved the problem of where to make a taxonomical cut. What was the appropriate unit of cognitive activity; was the sight (both how seeing worked biologically and what was being seen in the experimental results) composed of one distributed unit or several distinct components? There were secondary questions as well, including what was a scientific observation and what was an artifact (as the debate around dendrites shows), but the central quagmire was a visio-epistemological one. Are we seeing the same things? If so, are we making the same cuts on the same visual? As far as Ramón y Cajal and Golgi were concerned, the answer to the former was unclear but the latter was most definitely no (see Daston and Galison 2007).

And this question itself gets complicated in/after an era when microscopes were being constantly improved and adapted (Wilson 1995). In other words, the causal loops in this story do not flow unidirectionally; it is not just the case that scientific media were providing visualizations that were disagreed on by two competing scientists, or that proclivities and demands for certain visualizations were spurring the need for scientific media in one direction or another. The relationship between scientific media and scientific inscriptions is best understood as a Möbius strip running from and toward itself, the push and pull entangled being all intra-active (Barad 2007).

This reticular–neuron debate, it must be noted, has been well studied by science scholars. Cultural historian of science Laura Otis, for example, points out how the debate was essentially one about the epistemic status of a network; whether to see like a network or not was the question on the table, per Otis. She writes:

> Both those who argued for autonomous nerve cells (neuronists) and those who argued for a nerve net (reticularists) accused their opponents of "physiological prejudices." Rather than looking in a careful, unbiased way at the structures present, they claimed, the others were seeing imaginary structures that confirmed their theories about the way the nervous system functioned. Golgi accused the neuronists of physiological bias, and Cajal accused the reticularists of failing to free themselves from prevailing ideas, even of succumbing to hypnotic suggestion. These charges raise a vital question: what exactly determines what scientists see under their microscopes? (Otis 2001a)

Carrying forth this line of inquiry—what to do when two individuals just see different phenomenon under the same visual input—I would like to turn here, for an answer (or at least a hint), to an unlikely source but one that I think is at least generatively suggestive: the science fiction that Ramón y Cajal wrote under the pseudonym of Doctor Bacteria. In 1885 and 1886, when this debate was still raging and not yet completely settled, Ramón y Cajal took some time off from his scientific research to write science fiction stories, *Cuentos de vacaciones,* which were later published in 1905 (and much later translated into English as *Vacation Stories*), a year before he won the Nobel prize (ironically shared with Golgi). (I shall take up the issue of this temporal gap between controversy and publication later in this piece.) In his memoirs, Cajal is clear about his science fictional influences—Jules Verne makes a prominent

ce, for example—and he notes that he even wrote a com-
cience fictional novel that he lost during his military service
ain and Cuba (Otis 2007; see also 2001b).

onsidering the long history of lab-lit, or laboratory literature, we can see how and why the practice of science fictional writing by scientists is no great surprise or shock by itself (Pilkington and Pilkington 2019; Rohn 2005). Ramón y Cajal was, in fact, very self-consciously taking a "vacation" from his scientific pursuits when he penned these tales. But I want to point out here three critical elements that stand out in these tales specifically, and in Ramón y Cajal's practice more generally as a science fiction author.

First, it cannot be understated how radically Ramón y Cajal's project relied on a confirmation of a specific *way of seeing*. This, perhaps, is the most unsurprising of observations; known even today for his exquisite renderings, Ramón y Cajal's project, after all, stood atop visual observation—more specifically observation performed by a supposedly independent subject—as a specific scientific method. "Perceiving phenomena under the lens of the microscope and their subsequent artistic rendering placed [Ramón y Cajal] at the center of innovation, veneration, and obsolescence related to ways of seeing," writes Claudia Schaefer. "In a life that spanned many social and technological transitions, [Ramón y Cajal] also promoted and practiced photography, from collodion emulsions to portable cameras" (Schaefer 2014). This focus on vision, not distinct from the longer history of the visual in the sciences (one can certainly go back to Vesselius's anatomical renderings here, and maybe even further), nevertheless was also very visible in Ramón y Cajal's fiction. In one peculiar story, "The Corrected Pessimist," he imagines a scientist given the optic powers of magnification by several orders. As a result of this merger of the microscope with the scientific mind, Ramón y Cajal's protagonist ends up unnerved when he notes the minute details that he was never meant to see in paintings, in the airborne particles he breathes, or even in his love interest's face.

A quick rundown of the story is in order here.[6] The protag[illegible] scientist named Juan Fernández, has met with several failure[illegible] life: his academic pursuits have not yielded him a university ch[illegible] and his ability to one day marry his love-interest, Elvira, has come under doubt as his sadness and disappointment start affecting his social life. In a particularly angry moment, he is visited by a vision called the "spirit of science"—who remarks that "the philosopher calls me intuition; the scientist, fortunate coincidence; the artist, inspiration; the merchant and politician, luck"—and a lengthy dialogue between the two follows (131). It is here in a remarkable dialogic diegesis that Juan, the scientist, discusses and works out his theories about the world with a representative of that quasi-divine "world." This includes philosophical and theological problems (such as why does the "Supreme Cause" permit the "bloody struggle for life," and whether pain is an ecological necessity or merely a psychological effect) and scientific ones (such as do bacteria play any useful role in the larger biological system of our world). As the conversation moves forward, the protagonist wonders why he, a scientist who wants to know more about the world and improve it, cannot be granted the powers of optical magnification in his eyes. The spirit exasperatedly grants him the boon temporarily for a year, assuring that Juan will see it for a curse soon. And thereby begins Juan's tumble through the problems of seeing the world in magnification.

I want to now zoom out and put this tale in conversation with Ramón y Cajal's own life and science. My first maneuver here, again not uncommon in laboratory literature criticism, is to see Ramón y Cajal, someone who had lost academic chair *oposiciones,*[7] as a stand-in for Juan. Oscillating between the age of microscoping advancement on one hand and the perils of microscoping the whole world on the other, Ramón y Cajal (like his protagonist, Juan) had envisioned a study of brain that relied on vision, with the visual standing in for veracity in his scientific experiments and their subsequent renderings. It should be noted, then, that Juan does not use his magnified eyes solely as a scientist upon being granted the boon/curse; his eyes are merely a functional substitute for the microscope (which is to say he sees with his naked eye what anyone can see with a microscope) and would grant him no advantage *unless* he used them in conjunction with a microscope (thereby increasing his microscoping capabilities when compared to other scientists). In other words, the ocular, even when extraordinary, relies on optical media (Kittler 2010) for its scientific operations, both in practice and in narratives of the same. Furthermore, an important problem arises when Juan actually starts using, after initial disappointment and paranoia, his enhanced ocular-optical capacities for scientific practice. Confident that he would make "portentous discoveries" and "supreme conquests" to be recognized as "an extraordinary genius, an analytical demon, a monster of penetration, intuition, and logic"—all associations of an era of science that took pride in the quasi-colonial superiority of western European science and its breakthroughs—Juan sets to work (Ramón y Cajal 2001, 140). He first uses telescopes to resolve "the most arduous problems of planetary physics, chemistry and biology: the atmosphere of the moon, the habitability of Jupiter, the question of Martian canals, the chemical composition of stars, etc." Then, he turns to micrography and bacteriology, studying "diseases of unknown origin" and revealing "the ultramicroscopic germs of cowpox, smallpox, measles, syphilis, tumors." "But ah! These admirable findings [run] into one minor obstacle . . . No one believed them" (159). Juan, and by extension Ramón y Cajal, discovers what every scientist always already knows (though not always necessarily in a self-aware fashion) and what science studies would explicate in detail much later: science has always been a social form as much as a technical practice,[8] and its operations and relations only arise out of different kinds of consensuses. In highlighting these parallels between Ramón y Cajal's own scientific milieux and that of his narratological pursuits, I wish to focus attention on the role of Ramón y Cajal's storytelling for his—and, here I make one final proxy extension, his times'—science.

-cond, I would like to point out that science fiction, when taken as parascientific media, may also be understood as a medium that Ramón y Cajal was using to think through the issues—epistemic, visual, scientific, communicative—that are at the core of his work. It is, nevertheless, not a straightforward ploy to get scientific and other, more general publics on Ramón y Cajal's side. The canonical vocabulary in science studies for talking about phenomena of this sort is militaristic; Bruno Latour (1993), for example, talks of jostling, cajoling, and convincing allies and enemies, and of problems of translations that make skeptics and enemies into allies in moments of crises. In that vein, Ramón y Cajal could easily be construed as recruiting publics (Hauser 2022; Warner 2021) to his very visual cause of the neuronal doctrine. His extraordinarily detailed renderings even today tour the art circuit in exhibits (Newman, Araque, and Dubinsky 2017; Smith 2018) and evoke a social visuality; in some, the cells look like trees, while in others, forms such as weather systems show up. By identifying (with) Ramón y Cajal's freehand doodlings, the publics (both scientific and nonscientific) necessarily had to rely on their imagination and understanding of the world as it existed. And it could be argued that his science fiction must be read in the same way. Ramón y Cajal here was experimenting with these literary-scientific modes of thinking. His use of pseudonym (Ramón y Cajal calls himself "Doctor Bacteria" for the purposes of these stories) in an age of pathogenic fear and cholera should be understood as him building a paratextual apparatus of social acceptance around his narratives. One story, "For a Secret Defense, a Secret Revenge," in fact, features a bacteriologist gone rogue; this scientist protagonist suspects his wife of infidelity and infects her supposed lover, who also happens to be his own lab assistant, with bovine tuberculosis bacillus. He then waits to see if his wife gets infected, and "writes up his results in a bacteriological journal" (Ramón y Cajal 2001, 1–37); this text is later discovered by his wife and her lover—both removed from the action because of their infections—closing the loop between the scientific discoveries and personal vengeance. (To be fair, the story drags on for quite a bit and features yet another round of wildly unethical behavior by our dear scientist before the loop is fully closed, but I shall not you here with the details of all Ramón y Cajal thinks brilliant men can achieve.) It is noteworthy that all throughout these twists and turns, the protagonist's fame (and his scientific standing) remains front and center of the narrative and in fact provides the substrate for his personal machinations and his phallic follies. In collapsing Ramón y Cajal and Dr. Max v. Forschung (the aforementioned protagonist), one may note that social-scientific entanglements remain fluid and move freely between the stories and the science in the late nineteenth century. But this clear throughline—scientist publishes science fiction to cultivate public appeal—is complicated by the fact that Ramón y Cajal did not actually publish his stories initially. He was unsure about the suitability of such tales for a wider audience, and he waited for nearly two decades; many of his stories were apparently scandalous enough to have been lost or destroyed. Finally, he published these stories for the first time in 1905, a year before his Nobel win, right when his theories were starting to gain consensus in the scientific community. In other words, the gap in the publication history of Ramón y Cajal's science fiction is evidence that parascientific media here is not being used as an active, conscious public motivator for ally formation *during* a scientific controversy; if anything, it is being used in the aftermath of the controversy, and that too for stabilization and consolidation of the claims, not their initial dissemination. The publics being formed and sought here may then be seen as slightly distinct from the ones that will be formed in the case of parascientific media intervening in an ongoing debate, as Kaplan and Radin's formulation suggests. This gap also suggests the broader stakes of my argument here, since the writing of these stories during the debate offers us a location for working through the positions and arguments taken by a side *during* a scientific controversy, even if their publication *after* the debate was settled indicates a different modality of mediation. Quite simply, the stories, when they were being written, were a playground, or an experimental medium, for thinking.

The fact that science fiction is fundamentally a mode of scientific thinking pursued differently hints at the scale and stakes for these

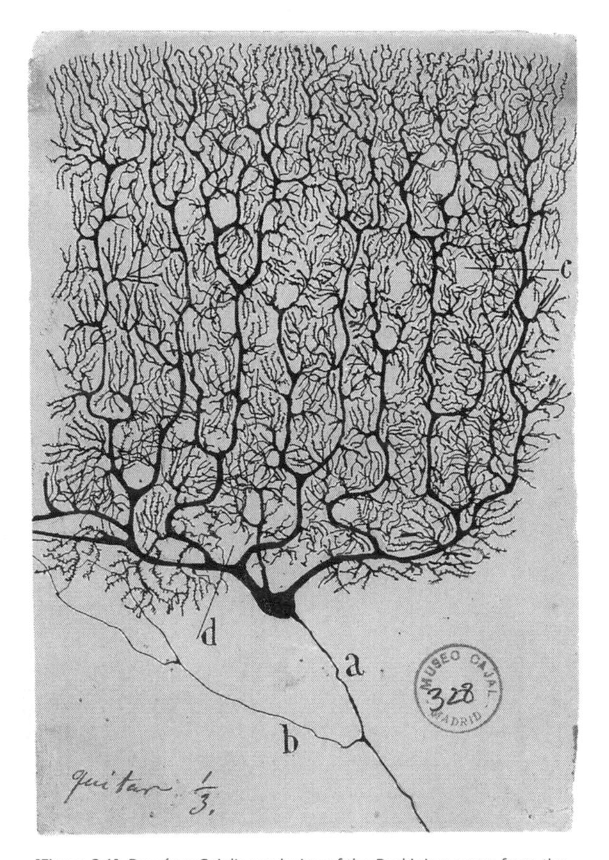


[Figure 2.1]. Ramón y Cajal's rendering of the Purkinje neuron from the human cerebellum that adorns book covers and gallery spaces. Source: Cajal Institute (CSIC), Madrid; https://commons.wikimedia.org/wiki/File:Cajal_-_a_purkinje_neuron_from_the_human_cerebellum.jpg.

operations. These epistemological stakes are not always about explicit social tactics and ally building, as may be the case in canonical Latourian readings, but are in fact embedded, unseen modes of thinking socially. Allow me to elaborate on the science fictional aspect of these tales to make my point. Science fiction as a mode of thinking, as the subdiscipline of literature and science reminds us, is foundationally experimental in nature (see Milburn 2010). The ability to spot the/a world in fiction is a hallmark of all science fiction, as scholars such as Darko Suvin have explored. In Suvin's argument, science fiction operates as a genre of cognitive estrangement, as it makes you think differently in a consistent fashion (Suvin 2005; see also Haraway 2013). Ramón y Cajal's ploy here, then, can only be in the most cynical reading understood as one of recruiting, or even solely post facto consolidating, this generic tendency of science fiction for public outreach. Ramón y Cajal did not, at least initially, as mentioned above, rush to publish his fictional work, even though privately he was distributing his stories. This push and pull—the use of a pseudonym but also the retroactive claiming of his own name, and the writing and distribution but not immediate publication—suggests a balance between the scientific and the parascientific that Ramón y Cajal was acutely aware of. This dynamic, I would like to argue, tells us something significant about the role of media such as science fiction in the controversies that Ramón y Cajal found himself in, controversies in which we find the idea of a neural network first appearing, as mentioned in the Introduction to this book. Ramón y Cajal's fictional tales, as pointed out in this chapter, were often reflecting on cognition and its relationship with visuality; at the same time, Ramón y Cajal's scientific practice was arguing for understanding interconnected individual cognitive units (neurons) as the core of cognition instead of a network-like web of brain matter, as Golgi had suggested.

These parascientific media, as the examples above demonstrate, are inscriptional markers of the scientist—and his publics, whether real or imagined, present or sought—working through their own thoughts and positions in alternative forms and fashions. In other words, science fiction is useful for Ramón y Cajal not necessarily, or at not least solely, because it allows him to reach other actors and convert them (or consolidate them) into allies for his theories and his visual mode of scientific demonstration, though that is certainly one way of understanding some variant of this story. More importantly, in my account, it provides him with a new set of media forms—narrative forms, plotting, character development, hypotheticals, novum, to name a few—that let him explicate, interrogate, contemplate, and refine his (socioscientific) thought. For the scientist and his readers, it is here the role of dialogic protestations with the "spirit of science" is made clear and it is here that the social reception of scientific renderings and journals is brought into light.

Science fiction is by no means the only medium in which this thought is fostered; another common genre for such action is science advice books, though I will also have more to say about some other media in the next section. Ramón y Cajal is also known for his *Advice for a Young Investigator,* which was first published in 1897 (Ramón y Cajal 2004). Full of anecdotal observations generalized into scientific advice and maxims, this scientific self-help genre (and I am certainly being a little facetious here in retroactively labeling it as such) is yet another location where thinking *through* a problem that animates scientific controversies (which is to say, processes, methods, and discoveries) is made visible and palpable in a mediatic form that is alternative to scientific notes and publications (which have been called by Rheinberger [2003] the scrips and scribbles of science). His advice book remains a bestseller more than a century later, and it continues to be given as a gift to recent scholars (as I personally found out upon attending a neuroscience graduation ceremony). Such practices further the methodological claims and myths generated by these genres and the scientists who use (and reuse) them. In other words, not only does the given (para)text suggest modes of medial thinking at work in Cajal's science fiction and advice writing, but it also clearly indicates the practice of inducing-and-consolidating publics toward his positions.

Insofar as claims about neurons and networks are claims about modes of thinking, Golgi and Ramón y Cajal were seeing the world *in* their diagrams. Here, the claims that the two were making on modes of knowledge were also claims about thought and, in turn, formal modes of making claims; this corresponds with Otis's argument about seeing networks in the world in the age of networking. This, of course, is not about right or wrong knowledge but about the apparatus around knowledge claims as they self-reflexively work in the world. In other words, Ramón y Cajal's project—making claims through speculation (*Vacation Stories*), didacticism (*Advice for a Young Investigator*), and visual rendering (diagrams and paintings)—was successful and stable not only because it recruited/retained certain kinds of publics to his side after a scientific controversy, with the promise that his claims—about neurons being the constitutive unit of thinking—saw in the world a mode of networked operation that was socially and rhetorically prevalent. The project allowed Ramón y Cajal to think through and with the problems at hand, finding the right rhetoric, the right techniques, and the right forms of articulation (and perhaps just the paratextual thrust in another sphere) that will make the scientific world run. Each time you study a new part of the organic world, such as the neuron, your claim to the neuronal doctrine gets strengthened; it would only be logical then, in this line of thinking, to argue that scientific claims are additive and networked, thereby necessitating parascientific media forms for their growth, and in some cases, succor.

## Scene Two: Educating Neural Subjects with Coursera

Our second scene of controversy finds us in the early twenty-first century, right around the thaw of what was dubbed an AI winter.[9] In this story, the identities of the protagonists of the two competing sides are slightly less clear just now; maybe we will get a Golgi and Ramón y Cajal–type caricature later as historiographers ponder these questions. Nevertheless, it was clear that artificial intelligence in late 2000s and early 2010s found itself exploring two possible directions. On the one hand, AI could have developed through a more sustained engagement of logical pathways (the symbolist model), and on the other, emergent phenomena, as encapsulated by artificial neural networks, could have had a resurgence (the connectionist paradigm). We now know that the connectionists with their neural networks won this particular battle for power, prestige, grants, and scientific bragging rights. But like our first scene, rearranging the spotlights *around* the purported scene—that of a bench last time but maybe an IDE here that is surrounded by windows[10] of YouTube videos, educational content, StackOverflow, Khan Academy videos, and Coursera modules—might help us figure out what was happening *around* the stage that usually demands our attention. Much like our first story, parascientific media in the tale that follows might recruit "allies" amid (and definitely after) controversy and debate, but more importantly, such media provide newer substrates for the scientific thinking and practice. These newer substrates are reflective grounds for thinking about the practices that make up (some of) the debate in the first place. I do not tell this, or the last, story to form direct causal lines between the use of parascientific media by one side and its eventual (and apparent) "victory," to use Latourian terminology again, but instead to point out how and where we can see streaks of epistemological utility and cognitive action *around* the scientific—or in this case technological, if not technoscientific—drama by itself.

Here I am following Andrew Ng, the British-Singaporean-American entrepreneur and scientist who has worked at Google, Baidu, and Stanford, among other locations. What is curious about Ng, who happens to be one of the highest-cited artificial neural network researchers in this recent AI boom, is that he also happens to be deeply dedicated to his pedagogical mission. He was adjuncting at Stanford when he started working for Google in 2011—more specifically, in the Google Brain project co-founded and, with Jeff Dean, led by him. Google Brain, as the name suggests, was a research team at Google that explicitly drew upon biological structures for advancing machine learning systems, indicating a techno-biological imbrication at the very level of nomenclature. (It would later, fittingly, be merged into Google Deep*Mind*.) Around the same time, Ng started working on pivoting the already-existing classes that he was teaching in Stanford to online modes of asynchronous teaching. Individual course websites—the "applied" version of the Stanford "Machine Learning" class (CS229a), followed by classes on databases and AI—went up starting October 2011. And by 2012, Coursera as a MOOC platform had been launched by Ng, his colleague Daphne Koller, and his overworked student team. (A story that Ng likes to "neither confirm nor deny" recounts one student refusing to give another a lift and as a result the said person choosing to work all night to put in features demanded by the ever-growing student population [Ng 2016].)

What is curious about Coursera is not just its business model (which is ostensibly about using the cultural capital of prestigious universities to democratize knowledge while handing out paid accreditations that, on paper, try to replace the costly university accreditations, one of the many models implicated in the neoliberalization of the university) or its platform-based approach, or even its popularity (popular courses regularly get more than a hundred thousand students each)[11] but more importantly, its entangled origins with Ng's research and advocacy of neural network paradigms. Ng himself, after all, often points out the formal similarities in how and when his two successful ventures became famous (Ng 2016).

Ng, it must be remembered, was the first prominent connectionist to advocate using graphics processing units (GPUs) in order to instantiate deep neural networks in hardware. This choice was initially controversial—GPUs are not built for AI applications, or at least were not built for that back in 2008 (Dhaliwal 2021)—but Ng had a big audience for his approach; his Stanford classes were regularly full and getting great press. *Forbes, New York Times,* and *Wired* had all profiled him. In these profiles, the boundary between scientific pedagogy, technological progress, and popular opinion is often blurred. Ng, as a boundary subject, moves between Stanford as a teaching institution, Stanford as a research location, Stanford as a credentialing service, and Google Brain as a private research laboratory freely in discourse and in the general public imagination.[12]

Coursera was drawing from Ng's own success as a teacher, since he had been firmly established as the realm priest for connectionist paradigms, when it initially started exhorting people to "Take the World's Best Courses, Online." The logic here came from the fact that Ng himself had after all offered one of these "best courses" in the world. The mythmaking around his courses, I should reiterate, was indistinguishable from the mythmaking around the connectionists. In the *Forbes* profile, even Marvin Minsky, a prominent symbolist from the previous AI wars, gets a mention as an antagonist, bringing Ng into relief as a genealogical successor to the connectionists who had lost their first battles but were now coming back to win the (AI) war.

I want to note two striking features of this nexus. First, we can easily see how the introduction of Coursera as an integral part of the technoscientific push-and-pull for AI prestige and funding not only integrates the Silicon Valley into the conventionally understood "scientific funding" model coming from the science foundations[13] (Coursera is venture capital [VC] funded and has not yet turned a total profit, as is common for such products of speculation) but also brings the technological problem of artificial intelligence "progress" into the realm of internet and visual culture. As the fortunes of Coursera and artificial neural networks were tied together,[14] the output from the technology was becoming more integrated with the internet visual culture. On the one hand, as I have discussed elsewhere, even the very presence of GPUs as infrastructural bases for the deployment of neural networks hints at something visual-cultural in the AI milieux (Dhaliwal 2021). On the other hand, we also see the medial form of Coursera—notes, lectures, videos, quizzes, and transcripts all collide in the visual frame—taking on the user interface (UI) expectations generally associated with visual internet media. Here, on the screen, the structure of the course, where modules always show up on the left as you listen to the lectures, validates the content within the video at the center. Coursera takes a classroom setting and converts the AI discourse in that space into a thoroughly visual cultural object, correspondingly also nudging the conversion of pedagogy into a neurally networked operation.

Second, the merger between educational media (Coursera) and scientific media (the inscriptions produced in the "lab") provides the latter with another location of work and experiment. It must be noted that for a problem as abstract,[15] ambiguous, and pervasive as "artificial intelligence," the need for a wide variety of instantiations as experimental setups has always been important. Scholars have pointed out the express connections between AI work and chess (Ensmenger 2012; Larson 2018), other games (S. Jones 2008; Hennig 2020; Skinner and Walmsley 2019; Milburn 2018), craftwork (Suchman and Trigg 1993), and language (Li and Mills 2019; Donahue 2022 and 2023). This, one might ascertain, is a concern inherent to AI research at large, insofar as everything and nothing can be socially coded as an intelligence problem (Dhaliwal 2023). But more importantly for our story, the use of educational media such as Coursera generates yet another arena for working *through,* if not working out, the conceptual and pragmatic conditions of artificiality and intelligence. The breaking down of famous AI experiments into modules and subsections—time and load management that all teachers think of—gives the bases for the structural and formal configuration of artificial neural network practices here.

There is already some scholarly consensus on the role played by these educational media. It has been pointed out by Luchs, Apprich, and Broersma (2023), for example, that educational courses offered by Google and IBM in these subfields establish corporate hegemony over technical education. In their words, such modes of online education help these companies recruit new AI talent and establish the dominance of their own "infrastructures and models." Though this observation also rings true in the case considered here, our story predates the contemporary Google education modules examined by Luchs et al. The desired dominance of certain "infrastructures and models" is further complicated in this tale by the fact that these models are not yet fully formed at this moment in the early 2010s. Before the dominance of technical models can be established, their inventiveness and utility needs to be confirmed in public discourse. And herein lies yet another function of these courses. It is no trivial matter to convert questions of artificial intelligence at large into questions and concerns around GPU-based neural networks more specifically, and Coursera helps Ng (and his peers) in this story to do just that. Thousands of students logging in to learn about "artificial intelligence" (who were first being taught to understand an apparently stable concept of *machine learning* and to then conflate it with *AI* and further with *neural networks,* as a quick glance at Coursera's webpages over time shows) were being also converted into "neural subjects," a question I return to later in this chapter (see also, Dhaliwal 2023; Suchman 2023). In this endeavor, converting principles, beliefs, and experimental systems into pedagogical forms also provided a pathway for thinking through those principles, beliefs, and experimental systems in the first place. Coursera, in other words, provides its own sandboxing media that all help connectionists think through and with their objects of study/analysis/promotion, and helps stabilize public knowledge into positions held by its creators.

Much like the first scene of this chapter, Coursera is not the only parascientific media at work in the case of artificial neural networks; the use of popular press and interviews, at least in the case of Ng, is yet another venue for scientific work *as* such. In this volume, you will also find examples of such engagement as taken up by a second-order collaborator of Ng, Geoffrey Hinton,[16] whom Lucy Suchman studies through his public outreach in the next chapter. And one can enlist (going back to the militaristic Latourian worldview) media forms as varied as online forums (StackOverflow and Quora Answers, for example), conference and summit presentations (from TED Talks to PowerPoint files; see Robles-Anderson and Svensson 2016; Cornfeld 2022; Mattern 2014), or campus forms to be filled as advisors and bureaucrats, and even grant writing done for agencies/foundations and business pitches for investors (both fundamentally speculative forms of writing) as parascientific media for thought where neural network controversies—and more importantly, their aftermath—find their stable configurations without the careful, conscious, or necessary knowledge of the actors therein.

Not all the *para-* prefixes in this story are equivalent; paratextual does not always correspond neatly to the parascientific, which is not the same as the para-academic. But all these formations work as mediational agents in scientific controversies such as the two around neural networks highlighted here. Pedagogy (in case of Coursera) and fictional outputs (in case of *Vacation Stories*) are all fundamentally cultural in nature and often go hand-in-hand. The two most prevalent epistemological forms of the content that this parascientific media peddle in are speculation (in case of Ramón y Cajal's science fiction and Ng's VC pitches) and didacticism (Ramón y Cajal's scientific self-help advice writing and Ng's Coursera platformization), though others can be imagined and perhaps imbricated within these quasi-dialectical modes. These relationships between parascientific media on one hand and the "core" scientific action on the other need not necessarily be causal, though that is always a possibility. Nevertheless, regardless of causalities and intentionalities, suggestive lines of interrelations between the two can lead us to see how scientists and technologists frequently seek to utilize media domains that often otherwise go undernoticed by STS scholars and even by scientists themselves. If parascientific media do not help solve controversies, that is no big problem for our actors at all, as long as the given media can help settle or renarrate stories retroactively in a certain way around/after controversies have been settled.

Here it is worth mentioning that parascientific media—precisely because they perform such an outsized role in helping stabilize and recruit postcontroversy publics—are also crucial cogs in the machines that form "neural subjects" (Bates and Bassiri 2016; Pelaprat and Hartouni 2011). "Neural subjects" can be very simply understood as a group of individuals who all comprehend their cognitive activity and even their subjectivity through neural frameworks; that is to say, somewhat crudely, that such subjects believe that neurons and their networks grant intelligence and cognition and, as a result, the biological basis of the very thinking firmament of human action. It follows, then, that neural subjects are also in the practice of understanding artificial neural networks as *the* model of computational intelligence and as a distinct (and valid) challenge to their own subjectivities. The subjects (and the publics they constitute) that coalesce around these ways of seeing, knowing, thinking, and doing start to agree, as a result, on the basic principles laid out in the approaches used by the proponents of artificial neural networks. Neural networks thus become the primary mode of artificially intelligent operations in popular discourse, and the publics start to blur the boundaries between the biological and the computational (Bruder 2019). In such a world, the travel of the semiotic matter between the neural, the networked, and the neurally networked becomes even more prevalent.[17] One of the questions on table, for Ramón y Cajal and Ng, is how to make their project intelligible as a part of knowledge creation. Both men tried to reshape a broader field and its practices, conflating neurons with intelligence, and neural networks with artificial intelligence.[18]

This redefinition of the field (by implying that artificial intelligence is to be primarily understood through artificial neural networks) by proxy attempts to render future controversies impossible, or at least unlikely, since the stabilization of the argument that connectionist approaches to AI are *the* correct way of doing AI closes off some possible directions of research and thought in the coming years. The result is a neural subject who understands their self, their world, and their self-worth through a neural lens (see also De Vos 2016). Such a subject formation allows for discourse to flow from Silicon Valley to students wanting better careers through to public news media and discursive formulations on social media. Such control of the discursive terrain is also a key element of contemporary techno-capitalist cultures at large (Dhaliwal 2023). Parascientific media lie right in this space where other discursive openings are discouraged, and publics are converted into believers of the discursive grounds provided by the proponents of one position within what once was a controversy. If one were to ask what exactly do "parascientific media" say to us, one possible answer can be that they make claims about the suitability of certain technoscientific positions. That suitability claim, ultimately, is the primary claim made by parascientific media, a claim that is in search of its audience. At the same time, the audience—by existing within a media ecology of consent and needing politico-economic benefits (like jobs in Silicon Valley) offered by certain technoscientific positions—is primed and ready for the claim. The loop is then closed.

One should, however, not mistake such an observation for a transhistorical fact of science. The reason I analyze two distinct time periods in this piece is not only because what at first glance seem two unlinked genealogies demonstrate, upon a closer inspection, a surprising consistency in their mediational worlds. But I bring them together also because these two nodes provide me with a location to consider the historical specificity of media forms that are being converted into parascientific media (be it science fiction or pedagogical digital media). Media change over time, and mediational possibilities as sociocultural aggregations change with them. Hand-drawn renderings of neurons and Coursera's user interface may be ostensibly visual, but they are visual in different senses. In the former, the visual is ocu-optical, and in the latter, it is visual-internet-(sub)cultural. Neural intelligence, in these two examples, looks very different when seen through the context of two different technoscientific regimes; synonymous with optical intelligence at the start of twentieth century, it becomes socioculturally visual intelligence at the start of the twenty-first. The fact that the most positive press Ng ever got was for generating what *Forbes* calls "cute" (in parentheses) cat images as a platonic form is, as I have mentioned elsewhere, the perfect demonstration of the visio-cultural entanglements of contemporary AI technologies, an example that can be further elucidated by putting it next to generative AI software today (Dhaliwal 2021). Nevertheless, the affordances of media forms across time shape the possibilities and restrictions granted to the science at hand (see Hayles 2004).

In other words, despite the shared formal modes of speculation and didacticism, there are enough historical differences between the parascientific media used by our actors to merit a periodizing analysis of the phenomenon. The more things change in the para-scientific medial world, the more—or less—they remain the same.

There is a further point to be made here about the discursive specificity of my remarks above. As I hinted earlier, neuralities—conceptions of how intelligence works, be it organic or inorganic—are always contextualized within social understandings of cognition. In the case of Ramón y Cajal, neural networks appear as technologized physiology, dependent on optical and cultural media. In the case of Ng and Hinton, neural networks come to be seen as physiologized technology, organic metaphors seeping into their technological instantiations. But in both cases, they appear to stand in, and comment upon, a socially acceptable (or unacceptable) way of thinking about thinking. Whether collective or singular, mechanical or organic, human or inhuman, mysterious or comprehensible, rule-based or arbitrary, intelligence in our story remains the sociocultural core that both sustains and constrains the technoscientific work around and in it. In other words, it is especially because intelligence, both "natural" and "artificial" (Daston 1994 and 2018), is such a social formation that we can, more acutely than elsewhere, see the mediatic formation in society that links "science" to the parascientific "world."

## Curtains: (Parascientifically) Medial Thinking

In closing, it is worth returning to *how* exactly might a media form let scientists think. Media forms and formats provide vital scaffolding for the content within (Sterne 2012, 1–31). In the case of both the neuron and the artificial neuron, the conversion (whether we call it translation, transcoding, or rendering) of a scientific concept/hypothesis/idea into parascientific formats requires work. Such work, often tangential to the "proper" scientific work, nevertheless necessitates cognitive and material labor for its success. Let me point out two stages of such work that are indistinguishable from medial thinking.

In the first stage, the scientist in question needs to take the scientific idea and put bounds around it. Anyone who has ever had to make PowerPoint slides—the default form of which often requires bulleting and listing and sequencing, even if that is not how discourse or scientific inscription functions in the conventional arenas—would know about the work involved in making something PowerPoint ready. Here, medial forms establish a restrictive relationship with science. A soft science fiction story cannot have a chemical formula inserted in it; it is a genre convention that limits what can be said or done.

In the second stage, the very same media form that constrained the scientist in the first stage allows the scientist to rethink and reorient his objects and conditions. A Coursera course framework might force the teacher, for example, to come up with more questions that students can be asked, thereby forcing a reflection on the relationship between givens and inferreds in the case of neural network–based machine learning. Renderings of the cell, as provided by Ramón y Cajal, allow him to exercise his creative faculties in selectively making visible and invisible all the apparatuses near a neuron. In either case, the second step is the opening of newer lines of thought.[19]

Along and between these two stages of media thinking, we find parascientific media doing its work. Thinking with such media, then, is no neutral task; each conversion and each vector of freedom brings with it its own cognitive possibilities. Such thinking, as practiced by scientists and technologists, is often understudied and undernoticed. It can be best understood as moments and spaces of thought where a scientific problem, often indistinguishable from a social problem, finds substrates outside of scientific inscription as properly understood. Depending on the properties of specific media in question, such thinking often allows or restricts scientists and technologists to work out, gather support for, stabilize after settlement, subjectivize, and quite literally reformulate their positions and concerns in symbolic vocabulary—lines, words, diagrams, narrative forms, PowerPoint presentations, user interfaces, et cetera—that would otherwise not be afforded by their "scientific" setup. Here, easy distinction between fact and fiction, and between fiction and nonfiction, becomes troubled, and one sees a deep infusion of the scientific with the rhetorical, the metaphorical, and the figurative.

The fact that parascientific media have their versions of thinking attached to them is all the more apparent in this story precisely because the characters in this story are neurons, organic and inorganic. The demand for spaces to think out the social, cultural, and philosophical problems of cognition and intelligence—followed by my fellow contributors in this volume as well—is in some ways inherent to the neuronal here. It may be better to visualize this as the networks neural networks make. The form of thinking embodied by the concept here requires expansion, and parascientific media just happen to be the objects and projects around the bench (and the IDE) that are at hand. The neural network then spreads out leisurely and occupies these spaces, but not before it itself starts to get affected by those spaces. All stabilizations create their own believers, and all believers generate their own doubters. Parascientific media, for that reason, are never scientific enough for some, but that doesn't stop neural networks from thinking with them. What, if not the demand for expansion, is at the core of the network form?

Such an expansion also demands a critical counter. If observation in the past eras came with its own set of media (see Crary 1990; Benjamin 2002), then observation in the twenty-first century, if the stories of neural networks are to be believed, requires a critical approach to all that lies around, all that usually escapes our scrutiny. Attentiveness must be reversed, and distraction must be heightened for the para to be really understood within, and more importantly, around the technoscientific setting. Science may be off duty sometimes, but such media and neural subjects who think are never not working.

# Notes

I am deeply indebted to my coauthors, Théo Lepage-Richer and Lucy Suchman, for their camaraderie, patience, and vital feedback, and for joining me on this adventure. I shall never be able to properly account for the support and hard work of our editor-in-chief, the indefatigable Théo Lepage-Richer (who I am honored to call a friend and am fortunate to have collaborated with), and the care and wisdom of the indomitable Lucy Suchman (whose brilliance, selflessness, and thoughtfulness taught me essential lessons in humility and clarity of thought). I am also grateful to Sahana Srinivasan for everything, including editorial advice, and to Mushu and Manjinder for all their support. Thanks also to the two generous anonymous peer reviewers; to Wendy Chun, Mercedes Bunz, and Timon Beyes, the series editors; and the staff at University of Minnesota Press and meson press for all their help.

1 For more on thinking about thinking, see Dumit 2004; M. Jones 2016.

2 See also Lepage-Richer's contribution in this volume.

3 For a few canonical takes on this method, see Kuhn 1970; Daston and Galison 2007; Shapin and Schaffer 1985; Marres 2015.

4 For more on controversy studies in science and technology studies (STS) today, see Marres 2007.

5 Alternate to standard forms of scientific markings, in this case, well studied by STS since Latour. See Latour 1987.

6 All quotes from stories are taken from Ramón y Cajal 2001.

7 These *oposiciones* were peculiar academic battles that decided who would get a job; you can think of campus visits with multiple candidates in the room at the same time but bloodier and more fierce.

8 After all, Ramón y Cajal had been on the front lines of the fight against epidemics and learned this firsthand. During his tenure as a professor at the University of Valencia, in mid-1880s, he had found that making government officials understand the disease was as difficult as, if not more so, than dealing with the bacteria itself. In 1885, "the Provincial Government of Zaragoza, in recognition of his labor during a cholera epidemic, awarded him with a modern Zeiss microscope" (Bentivoglio 1967; see also Ramón y Cajal 1989). I am thankful to Théo Lepage-Richer for pointing this out to me.

9 The histories of artificial intelligence are legion. At the risk of oversimplification, artificial intelligence within the American context has had two prominent eras of disinterest and lack of funding, dubbed "AI winters," with the first one roughly around 1970s and the second in the 1990s. The story here lies at the end of the second AI winter. The two primary camps that have historically disagreed over approaches to AI are the symbolists and connectionists. The symbolists believe that AI is best thought of as a collection of increasingly complex but nevertheless largely predetermined and coded rules, while the connectionists believe that artificial intelligence has to be emergent and that the system must discover its operational form from aggregated and connected data-elements (see Lepage-Richer 2020; Mendon-Plasek 2016; Agre 1997; Nilsson 2009). Here,

in late 2000s and early 2010s, we see the connectionists and the symbolists vie once again for the next round of funding from Silicon Valley, which had only just recovered from first the dot-com bubble and then the 2008 stock market crash.

10 Virtual windows today are, after all, mediators of visuality that can be traced back to the Renaissance perspective and architectural windows, as Friedberg 2009 has shown.

11 The formation of these neural subjects—not just as "allies" in the AI wars but also as future professionals and laypeople who understood their own cognitive entanglements with machines—out of a large number of students is a productive topic I return to later in this chapter.

12 Another way of understanding Ng here would be as a shared agent with "interactional expertise" (see Baird and Cohen 1999; Collins and Evans 2002; Galison 1997 and 2010).

13 Lepage-Richer explores this idea in detail elsewhere. See Lepage-Richer and McKelvey 2022.

14 Initially, all the top courses on Coursera were AI related (see Kosner 2013; Mendez 2022), indicating how the university replacement was only for the talent-starved coding-jobs market but not necessarily for, say, prestige-oriented liberal arts class-credentialing (see Bourdieu 1973 and 1987).

15 For more, see Lucy Suchman's contribution in this volume.

16 Here, by second-order collaboration, I refer to the fact that Ng and Hinton share collaborators and social circles, even if they have not worked together directly as often.

17 Lepage-Richer is an astute analyst of this phenomenon. See his contribution in this volume; also see Lepage-Richer 2020.

18 In fact, one of Ramón y Cajal's vacation stories, "The Accursed House," (re)uses tropes from the gothic to explicitly have the scientist-protagonist resolve the narrative by explaining the supposedly supernatural haunting of a house using *his* bacteriology (Ramón y Cajal 2001, 69–121).

19 This, it can be noted, is akin to Hans-Jörg Rheinberger's claims around epistemic things, though the objects in question, for me, are not primarily "scientific" (Rheinberger 1997; see also Rheinberger 2016).

## References


Agre, Philip. 1997. *Computation and Human Experience.* Cambridge: Cambridge University Press.

Baird, D., and Cohen, M. S. 1999. "Why Trade?" *Perspectives on Science* 7, no. 2: 231–54. https://doi:10.1162/posc.1999.7.2.231.

Barad, Karen. 2007. *Meeting the Universe Halfway: Quantum Physics and the Entanglement of Matter and Meaning.* Durham, N.C.: Duke University Press.

Bates, David, and Nima Bassiri. 2016. *Plasticity and Pathology: On the Formation of the Neural Subject.* New York: Fordham University Press.

Benjamin, Walter. 2002. *The Arcades Project.* Cambridge, Mass.: Harvard University Press.

Bentivoglio, Marina. 1967. “Life and Discoveries of Santiago Ramón y Cajal.” In *Nobel Lectures, Physiology or Medicine 1901–1921.* Amsterdam: Elsevier Publishing Company.

Bourdieu, Pierre. 1973. “Cultural Reproduction and Social Reproduction.” In *Knowledge, Education, and Cultural Change: Papers in the Sociology of Education,* ed. Richard Brown, 71–112. Milton Park, UK: Routledge.

Bourdieu, Pierre. 1987. *Distinction: A Social Critique of the Judgement of Taste.* Cambridge, Mass.: Harvard University Press.

Bruder, Johannes. 2019. *Cognitive Code: Post-Anthropocentric Intelligence and the Infrastructural Brain.* Montréal: McGill-Queen’s University Press.

Collins, H. M., and Evans, R., 2002. “The Third Wave of Science Studies: Studies of Expertise and Experience.” *Social Studies of Science* 32, no. 2: 235–96.

Cornfeld, Li. 2022. “Demo at the End of the World: Apocalypse Media and the Limits of Techno-Futurist Performance.” *Communication +1* 9, no. 2: 1–10.

Crary, Jonathan. 1990. *Techniques of the Observer: On Vision and Modernity in the Nineteenth Century.* Cambridge, Mass.: The MIT Press.

Daston, Lorraine. 1994. “Enlightenment Calculations.” *Critical Inquiry* 21, no. 1: 182–202.

Daston, Lorraine. 2018. “Calculation and the Division of Labor, 1750–1950.” *Bulletin of the German Historical Institute* 62:9–30.

Daston, Lorraine, and Peter Galison. 2007. *Objectivity.* New York: Verso.

De Vos, Jan. 2016. “The Death and the Resurrection of (Psy)Critique: The Case of Neuroeducation.” *Foundations of Science* 21, no. 1: 129–45.

Dhaliwal, Ranjodh Singh. 2021. “Rendering the Computer: A Political Diagrammatology of Technology.” PhD diss., University of California, Davis.

Dhaliwal, Ranjodh Singh. 2023. “What Do We Critique When We Critique Technology?” *American Literature* 95, no. 2 (June 1): 305–19.

Donahue, Evan. 2022. “Towards an Android Linguistics: Pragmatics, Reflection, and Creativity in Machine Language.” *Proceedings* 81, no. 1: 156.

Donahue, Evan. 2023. “All the Microworld’s a Stage: Realism in Interactive Fiction and Artificial Intelligence.” *American Literature* 95, no. 2: 229–54.

Dumit, Joseph. 2004. *Picturing Personhood: Brain Scans and Biomedical Identity.* Princeton, N.J.: Princeton University Press.

Ensmenger, Nathan. 2012. “Is Chess the Drosophila of Artificial Intelligence? A Social History of an Algorithm.” *Social Studies of Science* 42, no. 1: 5–30.

Friedberg, Anne. *The Virtual Window: From Alberti to Microsoft.* Cambridge, Mass.: MIT Press. 2009.

Galison, Peter. 1997. *Image and Logic: A Material Culture of Microphysics.* Chicago: University of Chicago Press.

Galison, Peter. 2010. “Trading with the Enemy.” *Trading Zones and Interactional Expertise: Creating New Kinds of Collaboration* 21, no. 1: 147–175.

Haraway, Donna. 2013. “SF: Science Fiction, Speculative Fabulation, String Figures, So Far.” *Ada: A Journal of Gender, New Media, and Technology* 3. Available at http: www .markfoster.net/dcf/speculative_fabulation.pdf.

Hauser, Gerard. 2022. *Vernacular Voices: The Rhetoric of Publics and Public Spheres.* 2nd ed. Columbia: University of South Carolina Press.

Hayles, N. Katherine. 2004. "Print Is Flat, Code Is Deep: The Importance of Media-Specific Analysis." *Poetics Today* 25, no. 1: 67–90.

Hennig, Martin. 2020. "Playing Intelligence: On Representations and Uses of Artificial Intelligence in Videogames." *NECSUS_European Journal of Media Studies* 9, no. 1: 151–71.

Huhtamo, Erkki, and Jussi Parikka. 2011. *Media Archaeology: Approaches, Applications, and Implications.* Berkeley: University of California Press.

Jones, Matthew. 2016. *Reckoning with Matter: Calculating Machines, Innovation, and Thinking about Thinking from Pascal to Babbage.* Chicago: University of Chicago Press.

Jones, Steven. 2008. *The Meaning of Video Games: Gaming and Textual Strategies.* Milton Park, UK: Routledge.

Kaplan, Sarah, and Joanna Radin. 2011. "Bounding an Emerging Technology: Para-Scientific Media and the Drexler-Smalley Debate about Nanotechnology." *Social Studies of Science* 41, no. 4: 457–85.

Kittler, Friedrich. 2010. *Optical Media.* Cambridge, UK: Polity Press.

Kosner, Anthony Wing. 2013. "Why Is Machine Learning (CS 229) the Most Popular Course at Stanford?" *Forbes*. https://www.forbes.com/sites/anthonykosner/2013/12/29/why-is-machine-learning-cs-229-the-most-popular-course-at-stanford/?sh=42a3433c55b7.

Kuhn, Thomas. 1970. *Structure of Scientific Revolutions.* Chicago: University of Chicago Press.

Larson, Max. 2018. "Optimizing Chess: Philology and Algorithmic Culture." *Diacritics* 46, no. 1: 30–53.

Latour, Bruno. 1986. "Visualization and Cognition." *Knowledge and Society* 6, no. 6: 1–40.

Latour, Bruno. 1987. *Science in Action: How to Follow Scientists and Engineers through Society.* Cambridge, Mass.: Harvard University Press.

Latour, Bruno. 1993. *The Pasteurization of France.* Cambridge, Mass.: Harvard University Press.

Lepage-Richer, Théo. 2020. "Adversariality in Machine Learning Systems: On Neural Networks and the Limits of Knowledge." In *The Cultural Life of Machine Learning: An Incursion into Critical AI Studies,* ed Jonathan Roberge and Michael Castelle, 197–225. London: Palgrave Macmillan.

Lepage-Richer, Théo, and Fenwick McKelvey. 2022. "States of Computing: On Government Organization and Artificial Intelligence in Canada." *Big Data & Society* 9, no. 2: 1–15.

Li, Xiaochang, and Mara Mills. 2019. "Vocal Features: From Voice Identification to Speech Recognition by Machine." *Technology and Culture* 60, no. 2: S129–60.

Luchs, Inga, Clemens Apprich, and Marcel Broersma. 2023. "Learning Machine Learning: On the Political Economy of Big Tech's Online AI Courses." *Big Data & Society* 10, no. 1: 1–12.

Marres, Noortje. 2007. "The Issues Deserve More Credit: Pragmatist Contributions to the Study of Public Involvement in Controversy." *Social Studies of Science* 37, no. 5: 759–80.

Marres, Noortje. 2015. "Why Map Issues? On Controversy Analysis as a Digital Method." *Science, Technology & Human Values* 40, no. 5: 655–86.

Mattern, Shannon. 2014. "Interfacing Urban Intelligence." *Places Journal* Accessed November 24, 2023. https://doi.org/10.22269/140428.

Mendez, Manoel Cortes. 2022. "Coursera Sunsets World's Most Popular Online Course." *The Report*. Accessed November 15, 2023. https://www.classcentral.com/report/coursera-sunsets-machine-learning/.

Mendon-Plasek, Aaron. 2016. "On the Cruelty of Really Writing a History of Machine Learning." *IEEE Annals of the History of Computing* 38, no. 4: 6–8.

Milburn, Colin. 2010. "Modifiable Futures Science Fiction at the Bench." *Isis* 101: 560–69.

Milburn, Colin. 2018. *Respawn: Gamers, Hackers, and Technogenic Life.* Durham, N.C.: Duke University Press.

Newman, Eric, Alfonso Araque, and Janet Dubinsky. 2017. *The Beautiful Brain: The Drawings of Santiago Ramón y Cajal.* New York: Abrams.

Ng, Andrew. 2016. "What Is the Most Interesting Story from the Early Days of Coursera?" *Quora*. https://www.quora.com/What-are-some-good-stories-from-the-early-days-of-Coursera.

Nilsson, Nils. 2009. *The Quest for Artificial Intelligence: A History of Ideas and Achievements.* Cambridge: Cambridge University Press.

Otis, Laura. 2001a. *Networking: Communicating with Bodies and Machines in the Nineteenth Century.* Ann Arbor: University of Michigan Press.

Otis, Laura. 2001b. "Ramón y Cajal, a Pioneer in Science Fiction." *International Microbiology* 4, no. 3: 175–78.

Otis, Laura. 2007. "Dr. Bacteria: The Strange Science Fiction of Santiago Ramón y Cajal." *LabLit.* Accessed November 23, 2023. http://www.lablit.com/article/226.

Pelaprat, Etienne, and Valerie Hartouni. 2011. "The Neural Subject in Popular Culture and the End of Life." *Configurations* 19, no. 3: 385–406.

Pilkington, Olga, and Ace Pilkington, eds. 2019. *Lab Lit: Exploring Literary and Cultural Representations of Science.* Lanham, Md.: Rowman & Littlefield.

Ramón y Cajal, Santiago. 1989. *Recollections of My Life.* Cambridge, Mass.: MIT Press.

Ramón y Cajal, Santiago. 2001. *Vacation Stories: Five Science Fiction Tales,* ed. Laura Otis. Urbana: University of Illinois Press.

Ramón y Cajal, Santiago. 2004. *Advice for a Young Investigator.* Cambridge, Mass.: MIT Press.

Rheinberger, Hans-Jörg. 1997. *Toward a History of Epistemic Things: Synthesizing Proteins in the Test Tube.* Stanford, Calif.: Stanford University Press.

Rheinberger, Hans-Jörg. 2003. "Scrips and Scribbles." *MLN* 118, no. 3: 622–36.

Rheinberger, Hans-Jörg. 2016. "Afterword: Instruments as Media, Media as Instruments." *Studies in History and Philosophy of Science. Part C, Studies in History and Philosophy of Biological and Biomedical Sciences* 57: 161–62. https://doi.org/10.1016/j.shpsc.2016.03.002.

Robles-Anderson, Erica, and Patrik Svensson. 2016. "'One Damn Slide after Another': PowerPoint at Every Occasion for Speech." *Computational Culture* 5:1–38.

Rohn, Jennifer. 2005. "What Is Lab Lit (the Genre)?: Boffins Are So Last Century—Let's See Some Real Scientists for a Change." *LabLit.* Accessed November 23, 2023. http://www.lablit.com/article/3.

Schaefer, Claudia. 2014. *Lens, Laboratory, Landscape: Observing Modern Spain.* Albany: State University of New York Press.

Shapin, Steven, and Simon Schaffer. 1985. *Leviathan and the Air-Pump: Hobbes, Boyle, and the Experimental Life.* Princeton, N.J.: Princeton University Press.

Skinner, Geoff, and Toby Walmsley. 2019. "Artificial Intelligence and Deep Learning in Video Games: A Brief Review." *IEEE 4th International Conference on Computer and Communication Systems (ICCCS),* 404–8.

Smith, Roberta. 2018. "A Deep Dive Into the Brain, Hand-Drawn by the Father of Neuroscience." *New York Times,* January 18, 2018. https://www.nytimes.com/2018/01/18/arts/design/brain-neuroscience-santiago-ramon-y-cajal-grey-gallery.html.

Sterne, Jonathan. 2012. *MP3: The Meaning of a Format.* Durham, N.C.: Duke University Press.

Suchman, Lucy, and Randall Trigg. 1993. "Artificial Intelligence as Craftwork." In *Understanding Practice: Perspectives on Activity and Context,* ed. Seth Chaiklin and Jean Lave, 144–78. Cambridge, Mass.: Cambridge University Press.

Suchman, Lucy. 2023. "The Uncontroversial 'Thingness' of AI." *Big Data & Society* 10, no. 2. https://doi.org/10.1177/20539517231206794.

Suvin, Darko. 2005. "Estrangement and Cognition." In *Speculations on Speculation: Theories of Science Fiction, ed. James Gunn and Matthew Candelaria,* 23–35. *Lanham, Maryland: The Scarecrow Press, Inc.*.

Warner, Michael. 2021. *Publics and Counterpublics.* Princeton, N.J.: Princeton University Press.

Wilson, Catherine. 1995. *The Invisible World.* Princeton, N.J.: Princeton University Press.

[ 3 ]

# The Neural Network at Its Limits

**Lucy Suchman**

The connections between neural networks read as technologized physiology or as physiologized technology, depending on the direction of analogic travel, signal the longer history of traffic across the disciplinary boundaries of life and systems sciences (see Dhaliwal, this volume). Recognition of those connections is not the endpoint of inquiry, however, but the starting place. This chapter begins from that place to revisit the question of how sameness/difference is made between the brain and computer, not to resolve the question metaphysically or philosophically but rather to see how analogic thinking informs the projects of neuroscience and of computational neural networks, and where it breaks. I follow two threads in the rhetorical fabric of the neural network: the entanglements of technological analogy and physiological modeling in the work of neuroscience, and the retreat from model to inspiration when the alignment of brain and computer breaks in the technical work of computer science. I consider what the brain/computer analogy enables in the fields of neuroscience and computational neural networks, and what we might learn by foregrounding the moves that the protagonists of this story make when they reach the analogy's limits. More specifically, I examine how the brain/computer analogy breaks differently for feminist neuroscientist Gillian Einstein (who follows her research problem from the brain to sexed/gendered bodies and their worlds) than it does for neural networks researcher Geoffrey Hinton (who is committed to sustaining the analogy, even as he acknowledges its limits).

My argument in brief is that the enduring commitment that informs the project of computational neural networks—and its embrace by computational neuroscience—is to cognitivism. The sense of cognitivism in this context is a theory of intelligence based in a correspondence between mental representations formed in the brain/mind and a world taken to stand outside of it. Located inside the bounds of the skull (Star 1992) the brain/mind is receptive to its surrounding world in the form of inputs (rendered for purposes of processing either symbolically or statistically) and responsive through its outputs. The literature from recent anthropology, science and technology studies, feminist theory, and kindred fields critiquing cognitivism and articulating its alternatives is by now extensive.[1] Relevant arguments center on the inseparability of cognition from the lived experience of embodied persons-in-relation, with one another and with culturally and historically constituted social and material worlds. In these theorizations and associated empirical studies, the intelligibility and significance of social/material worlds is variously reproduced and transformed in practice, rather than given to individual brains/minds. The politics of practice, in the sense of how social ordering is enacted and with what differential consequences, are fundamental.

The cognitivist frame systematically relegates all that doesn't fit to the status of epiphenomena beyond the bounds of the science.[2] The commitment is not simply programmatic, in other words, but is what enables the perpetuation of the analogy of reasoning to computation. Sailing within the winds of cognitivism has led to a tacking back and forth between a logic or rule-based symbolic approach, regarded as "abstract reasoning," and a statistically based approach, figured as "deep learning." Posited as a remedy to the limits of the symbolic approach, the statistical approach now begins to show its limits. As recognition of those limits grows, the only course available within the closed world of cognitivism is a partial return to the symbolic, in the form of an imagined new symbolic/statistical hybrid.[3]

This vaguely articulated synthesis promises a solution that enables practitioners to stay within the boundaries of the cognitivist frame. In agonistic dialogue with these developments are a set of critical engagements that insist on attention to the different material and cultural histories of organic and artificial intelligences, where intelligence is understood not as brain-centered cognitive functions but as practices enlivened in and through ongoing relations of collective world-making, for better and worse.

## Articulating the Analogy

The figure of the neuron has been extensively traced within science and technology studies (STS), following its attachment as a prefix to a range of fields from neuroeconomics (Schüll and Zaloom 2011) to neurocultures (Vidal and Ortega 2011). Brosnan and Michael (2014, 681) observe that the designation by the United States Congress of the 1990s as the "decade of the brain," as part of a wider cerebral turn among science and technology funders internationally, resulted in the channeling of resources into neuroscience research. Their study of what they characterize as the enactment of the "neuro" in a neuroscience laboratory in the UK builds upon literature in the sociology of expectations (Brown and Michael 2003), attuned to the performative effects of promissory rhetoric in the institutionalization of technoscientific projects. More specifically, they analyze the promise of translation despite systemic boundaries between lab work and its clinical application, and the multiplicity of the neuro's enactment within as well as across those boundaries. The neuron in this context, they propose, doesn't so much cohere spatially as an entity as it is made to adhere temporally to the imagined future that will translate its materializations in the lab into efficacious interventions in the clinic.[4]

Another line of analysis within STS examines the biological reductionism of neuroscience (Dumit 2004; Martin 2004). Vidal (2009) traces the genealogy of the "cerebral subject" in its positing of the brain as the essential organ of personhood, with the body as its vessel. This brain-centrism, he argues, is central to the figuration of modern humanity since the seventeenth century, based in individualism and a self-consciousness separable from body and world. "The idea that 'we are our brains,'" Vidal observes, "is not a corollary of neuroscientific advances, but a prerequisite of neuroscientific investigation" (2009, 7).[5] Debates about mind as reducible to brain or as emergent are longstanding, however novel their manifestations in contemporary (computational) neuroscience. Moreover, while the object of laboratory work is the organic matter of biology, laboratory work is equally involved in practices of abstraction, as "the squishy stuff of the brain becomes a subject of graphic comparison, sequential analysis, numerical measure, and statistical summary" (Lynch 1988, 273). Quantification and its associated denaturing smooth the path for the neuron's travel from the biological laboratory to the research "laboratory" of computational neuroscience, with its own promise of translation from research to application.[6]

The limits of the neural analogy are less important in the case of artificial intelligence than the analogy's power as what Dennett (2013) first named an "intuition pump" for technical projects. Gary Marcus, a sympathetic critic of the project of computational neural networks, insists that the aim should not be for machines to literally replicate the human brain (whatever a literal replication of organic matter in machinery could mean). After all, he observes, the human brain is "deeply error prone, and far from perfect" (2018, 21), problems that by implication might be avoided in the design of computational machines. Yet there are, Marcus acknowledges, many areas in which the human retains an advantage. In a passage exemplifying a biotechnical imaginary that at once naturalizes the artificial and posits the organic as always already technological, Marcus suggests that "A good starting point might be to first try to understand the innate machinery in human minds, as a source of hypotheses into mechanisms that might be valuable in developing artificial intelligences" (2018, 21). In this time of computational ascendance, it is hardly surprising that the "innate machinery" of the brain would be seen as "a broad array of reusable computational primitives—elementary units of processing akin to sets of basic instructions in a microprocessor—perhaps wired together in parallel, as in the reconfigurable integrated circuit type known as the field-programmable gate array" (Marcus, Marblestone, and Dean 2014). While seeming to reject the binary of nature versus culture, these are rhetorical moves that naturalize the technological, rather than articulating either differences or relations between the organic and the machinic. The imaginary of the wiring diagram as a network exploits that term's etymology as naming an "open textile fabric tied or woven with a mesh for catching fish, birds, or wild animals alive," along with its expansion in the nineeteenth century to reference "any complex, interlocking system."[7] Once rematerialized as canals, railways, and other infrastructures for transport, the figure of the network is well positioned for its translation into mid-twentieth-century technologies of communications and information, along with the installation of calculation as a universal process, and the attendant erasure of the specificity of constitutive entities and relations.[8] The neuron can then be figured as a logic gate that determines the operation of synaptic connections (see Halpern 2022, 335).

With this backstory in mind, I consider how the analogy of technology and physiology mediates the project of computational neural networks and laboratory neuroscience, exemplified in the narratives that two contemporary practitioners offer regarding the logics and trajectories of their respective techno/scientific practices. As indicated, my focus is on how each conceptualizes their field of experimentation, and what direction their projects take as they encounter that field's methodological limits.

## The Neural Network in the Work of Computer Scientist Geoffrey Hinton

Geoffrey Hinton is widely recognized as a founder and leading researcher in the subfield of artificial intelligence devoted to computational neural networks, and more specifically to so-called convolutional neural networks or deep learning (see Lepage-Richer, this volume). An emeritus Distinguished Professor at the University of Toronto, recipient of prestigious awards, and former vice president and research fellow at Google, Hinton's aim "is to discover a learning procedure that is efficient at finding complex structure in large, high-dimensional datasets and to show that this is how the brain learns to see."[9] With degrees in experimental psychology and artificial intelligence, Hinton sits at the intersection of two laboratory-based approaches to human cognition.

In a conversation with then *Wired* editor-in-chief Nick Thompson (2019), Hinton offers this explanation of a neural network:

> **GH:** You have relatively simple processing elements that are very loosely models of neurons. They have connections coming in, each connection has a weight on it, and that weight can be changed through learning. And what a neuron does is take the activities on the connections times the weights, adds them all up, and then decides whether to send an output. If it gets a big enough sum, it sends an output. If the sum is negative, it doesn't send anything. That's about it. And all you have to do is just wire up a gazillion of those with a gazillion squared weights, and just figure out how to change the weights, and it'll do anything. It's just a question of how you change the weights.

This response to Thompson's request to explain what neural networks are exemplifies the slippery rhetorics of biotechnical translation. Implicitly taken as a question about computational neural networks, the answer begins with reference to "processing elements that are very loosely models of neurons." The gesture here toward computational entities as models of organic ones is qualified to suggest that the relation is more analogy than strong claim. The connections between elements have "weights," a familiar term in computational vernacular that refers to a set of mathematized values, which, Hinton explains, change through "learning." Here a process associated with organic life is used as a technical term referring to the computational adjustment of values based on better and worse results according to a prespecified outcome. The "neuron" reappears in the next sentence as an agent that "decides," based on calculations (over processes now rendered as "activities"), between binary alternatives. The power, Hinton explains, comes from a combination of the number of such processing elements and the techniques for manipulating the numerical values. The sleight of what is necessarily a human hand appears in the figure of the "you" who works out how to adjust those weights, a momentary shift in figure and ground from system to programmer that at once effects a difference between them and reveals their interdependence.[10]

Thompson follows with the question: "When did you come to understand that this was an approximate representation of how the brain works?" to which Hinton responds:

> **GH:** Oh, it was always designed as that. It was designed to be like how the brain works.
>
> **NT:** So at some point in your career, you start to understand how the brain works. Maybe it was when you were 12; maybe it was when you were 25. When do you make the decision that you will try to model computers after the brain?
>
> **GH:** Sort of right away. That was the whole point of it. The whole idea was to have a learning device that learns like the brain, like people think the brain learns, by changing connection strings.

While Thompson's question suggests that the computational neural network offers at least an approximate model for the workings of the brain, Hinton's response flips that relationship to position the brain as a model for the computational neural network. It is this inversion that enables the shift from model to inspiration, as when pressed by Thompson later in the interview on what is clearly a computational technique that deviates from any processes evident in the brain. Hinton demurs:

> **GH:** I'm not doing computational neuroscience. I'm not trying to make a model of how the brain works. I'm looking at the brain and saying, "This thing works, and if we want to make something else that works, we should sort of look to it for inspiration." So this is neuro-inspired, not a neural model. The whole model, the neurons we use, they're inspired by the fact that neurons have a lot of connections, and they change the strengths.

This reassertion of the difference between brains and computers belies Hinton's earlier statement in the same conversation that "we are neural nets. Anything we can do they can do." It is the polysemy of the term "neuron," referring alternately to the processing elements of the computational neural network and to the physiological entities that are the objects of laboratory neurosciences, that enables Hinton's statements about brain–computer relations to shift seamlessly between analogy and identity, inspiration, and model.[11]

As the conversation moves onto the topics of consciousness, learning, and finally Hinton's "four theories" of dreaming, things become increasingly ungrounded. With respect to the educational implications of neural networks Hinton opines:

> And we know now . . . you can just put in random parameters and learn everything. If we really understand what's going on, we should be able to make things like education work better. And I think we will. It will be very odd if you could finally understand what's going on in your brain and how it learns, and not be able to adapt the environment so you can learn better . . . And once [computational] assistants can really understand conversations, assistants can have conversations with kids and educate them.

The figure of learning is ubiquitous in both symbolic AI and neural networks discourse, traceable to the original Dartmouth Summer

Research Project proposal (McCarthy et al. 1955, 1) "to proceed on the basis of the conjecture that every aspect of learning or any other feature of intelligence can in principle be so precisely described that a machine can be made to simulate it." Conceptualized as a process located inside the brain, learning here is serviced by an environment that needs to be represented or modeled before it can be registered. It follows that it should be possible to reengineer the environment to optimize its alignment with the brain's processing requirements. Defined technically, learning in the context of neural networks references optimization of the computational processes required to produce a "correct" output, where the latter is a function of determinations made by humans considered to have relevant knowledge in a target domain. But in its capacities as a more floating signifier, learning stands as the holy grail for so-called artificial general intelligence, or what Marcus (2018) characterizes as "a human-like flexibility in solving unfamiliar problems."

In his critical review of progress in so-called deep learning, "a statistical technique for classifying patterns, based on sample data, using neural networks with multiple layers," Marcus (2018) identifies what he names "10 challenges" to the field. First among these is reliance on large amounts of data, necessitated by what he characterizes as the inability of neural networks to grasp abstract relationships. As an example, Marcus points to the ease with which his human readers, presented with the concept of a "schmister" defined as a sister over the age of ten but under the age of twenty-one, can identify whether they themselves have a schmister. "In learning what a schmister is," he writes, "in this case through explicit definition, you rely not on hundreds or thousands or millions of training examples, but on a capacity to represent abstract relationships between algebra-like variables" (2018, 6). Granted that our ability to make sense of the concept of "schmister" is not based on thousands or millions of training examples, just how, we might ask, is it a capacity to represent abstract relationships? This reader, in any case, thinks immediately of my own very lively and embodied sisters on hearing that word. And in what sense is this a relationship between algebra-like variables, beyond the numbers ten and twenty-one, unless the words sister and schmister are treated as associated text strings and not associated persons? The idea of abstract in this context already presupposes that knowledge equals the correct classification of instances into categories, ignoring other modes and references, not least to lived relations of kinship.

Marcus observes that the "deep" in deep learning refers not to profundity but rather to the number of middle-level or hidden layers of computational processes (introduced for greater efficiency). The actual shallowness of neural networks is a factor in their widely cited brittleness, which leads to some acknowledgment that they do not in fact engage in learning in anything like the human sense. Marcus (2018, 8) points out that the neural network trained to play the video game Breakout, which famously "realized that digging a tunnel through the wall is the most effective technique to beat the game," did not just fail to observe the rules. Rather, he notes, the system has no perception of digging, tunnels, walls, games, or rules, but only of pixels mathematized in such a way that, given a predetermined objective, it deploys statistical techniques to optimize the likelihood of that output on any given round. Moreover, neural networks have no way of dealing with what Marcus characterizes as "prior knowledge" (often referred to as "common sense"), a problem that cognitivists trace back to the lack of "abstract concepts."[12] Read more broadly, neural networks have no qualitative understanding of entities and their relations, an unsolved problem for AI that now leads Marcus to suggest the need for "hybrid models" and at least a partial return to representational or symbolic processes (Marcus 2018, 20). The premise that abstractions can be represented by symbols only reframes the problem as one of reference, however, while continuing to beg the question of what Lave (2011, 115) calls "knowledgeability in practice," including the inseparability of learning from the concerns of everyday life.

Most relevant for the concerns of this essay, Marcus points to what he describes as a "culture in machine learning that emphasizes competition on problems that are inherently self-contained" (2018, 12). This reliance on closed worlds sidesteps the recurring and unsolved problems arising from the limits of the computational sensorium. Neural networks require for their operation a dataset of machine-readable inputs amenable to analysis for statistical correlations that fit a set of predetermined outputs. Quantitative advances in the scale of datasets and the speed of processing do nothing to address the infamous difference between correlation and causation or address the place of causation in our reasoning about indeterminate relations and effects. Framed in terms of the brittleness or narrowness of computational neural networks, prevailing discourses implicitly reference a transfer model of learning that has been thoroughly critiqued by scholars like Lave (1988, 2011). Marcus characterizes this as "the problem of generalizing outside the training space," but Lave (2011, 155) rather draws on Stuart Hall's figure of "rising to the concrete" (2003, 131) to name the ability to bring generalized theory into relation with the specificities of ongoing, cultural/historical worlds. On this understanding, the reliance of computational neural networks on closed worlds is not just a symptom of their inability to deal with "novelty" but of a much more fundamental difference between computation and knowing in/as practice.

## Transboundary Explorations in the Neuroscience Research of Gillian Einstein

The limits of computational neural networks are managed through the effective closure, for practical purposes, of the worlds in which they are designed to operate. The efficacy of systems is measured through their performance on a small set of industry-standard "benchmark" tasks and registered as relative scores on competitive "leader boards" (Bender et al. 2021, 618; Inioluwa et al 2021). The promise of translation from laboratory to application that underwrites investments in computational neural networks is sustained through a combination of massive datasets and associated compute power, along with the overrepresentation of the scope of resulting capabilities. Most salient to this discussion, as we have seen in the case of Geoffrey Hinton, the strategic ambiguity of claims for the brain as a model for computational neuroscience enables retreat to the brain as an inspiration in the face of untranslatable differences between organisms and machines.

The figure of the neural network in the biological neurosciences is similarly inflected by twentieth-century technological imaginaries, and within the confines of the laboratory the neurosciences operate according to regimes of experimentation that offer another form of self-referential closure.[13] On the margins of mainstream neuroscience, however, an alternative project of translation is underway that builds on feminist critiques of heteronormative figurations of brains, bodies, and worlds to articulate a relational ontology of biological difference (Haraway 1989, Martin 1991, Fausto-Sterling 2012). To explore the implications of that project for the case of neural networks, I turn to the work of feminist neuroscientist Gillian Einstein, head of the Einstein Laboratory for Cognitive Neuroscience, Gender and Health at the University of Toronto.[14] In a paper titled "Situated Neuroscience: Exploring Biologies of Diversity," Einstein describes her project as one of "research into the nervous system that would give voice to areas of research previously silenced, uncover pockets of ignorance—not just 'knowledge gaps'—[and] turn expectations about the essentialism of biology on its head" (2012, 150).

To concretize what that could mean, Einstein reflects on her investigation of the neurobiological effects of the traditional Northeast and West African practice of female genital cutting (FGC). These effects, she hypothesized at the outset of her inquiry, might be more broadly constitutive of associated "normal/desirable" women's bodies:

> the result of the involvement of the central nervous system (CNS) would be to embody the tradition affecting the

> way women with FGC walked, carried themselves, and generally, experienced the world through their bodies thus, in effect, embodying their culture. I wondered specifically if the purpose of the tradition was to instantiate a corporeal difference in the CNS between male and female that wasn't present without the procedure. (151)

To pursue this hypothesis, Einstein realized, would require a series of extensions to the prevailing methods of experimental neuroscience. To begin with, there were no existing investigations of FGC that traced its effects neurobiologically. This "pocket of ignorance" in her view is sustained by the seventeenth-century model of the body as a machine with independent parts or systems (158). Within this model, she observes,

> the brain still sits privileged atop our polarized body with other body systems arrayed like arms, legs, and trunk on a marionette's strings—to be pulled and moved by the brain. Information comes in. The brain processes it. An action is generated and then carried out by the peripheral nervous system. The rest of the body responds. On this view, the brain is the CEO of the body. Perhaps because of this the body itself has not been considered knowledgeable and thus, has not been thought to have its own narrative. (159)

In contrast, Einstein insists that

> the brain isn't the only nervous system the body has. Other nervous systems are hard at work interacting and being affected by the rest of the body. The spinal circuits and the peripheral nervous system—nerves, receptors, and far-flung neurons—as in the retina, dorsal column nuclei and enteric nervous system—all contribute to what the cerebral cortex "knows" . . . This underscores the point that body, brain, and society are in a reciprocal relationship mutually affecting each other . . . the world writes on the whole body. (160)[15]

Beyond a commitment to tracing whole-body connections, then, Einstein's hypothesis required a more radical expansion of her methods, beyond bodies to worlds. This took her outside the laboratory into a collaboration with fourteen Somali-Canadian immigrants to Toronto (she emphasizes that they were positioned in the research as colleagues rather than subjects), in whose native country the practice of FGC still affected 98 percent of women at the time of her study. Einstein's collaborators, she is careful to point out, are not meant to be representative of all Somali women, or women who have experienced FGC: "Most were abroad visiting, studying, or working when the war broke out [in the late 1980s and early 1990s], and they simply never went home. They are healthy, engaged, energetic women with a particular sense of their place in the world" (163). And just as there is no singular Somali culture or category of woman, she emphasizes, there is a multiplicity of FGC.

The brain played an important role in the study, not only insofar as Einstein was interested in brain/body/world interconnections but also methodologically in the openings that a shift to neurobiology afforded in her conversations with her interlocutors. As Einstein explains:

> I was able to start out the conversation with each woman by saying that I was not interested in her genitals; I was interested in her brain. Redirecting the questions from the genitals—a site of silence in cultures practicing FGC—to the brain—a site not previously considered but privileged in the popular imaginary, allowed participants to talk about their circumcision as well as placing the topic in what for them was a respectful space. (161)

The results of the study confirmed FGC's whole-body effects for these women, including those expected in the initial hypothesis (Perović et al. 2021). At the same time, the study participants did not see themselves as disabled or unwell, and some expressed pride in what they had endured. While Einstein is mindful of the small size and specificity of her study population, she underscores its significance as further evidence for the inseparability of the brain from the body's multiple and interconnected nervous systems, and the inseparability of "pain" in both its measurable and its experiential forms from neurobiology's cultural embodiment.

More generally Einstein insists that the body has no independent parts, and rather than the privileged brain "the nervous system is an integrator of and integrated with the entire body and the world . . . Thus, a practice that affects one part of the body will be owned by the entire body or embodied through the interconnections of all body systems and the environment" (2012, 158). Einstein expresses her "love" for the endocrine system, observing that through the medium of the blood it works as "a huge unifying effector; people call it a modulator, but if you were really committed you could say that it was the starter, not the modulator. You could take any part of the body that's been studied as a system unto itself and realize that it's a modulator for the entire body and modulated by it."[16] Breakdown, moreover, is an opening to new knowledge: while the "small lie" that systems are distinct may have some predictive power, Einstein observes, "where it doesn't work is a good sign of where things are really interconnected" (Interview, May 2022).

Einstein conceptualizes the brain in terms of neural "circuitry," but for her the connections not only are profoundly and complexly embodied but also change with experiences not reducible to weighted inputs (2012, 162). Asked to talk about the limits of the neural network analogy in computational neuroscience, Einstein points to the premise that the top layer of the computational neural network is analogized to input receptors, while the middle layers determine, through back propagation and in unaccountable ways, the system's output. Neuroscientists, in contrast, care about what is in between:

> In the brain, a network consists of many neurons in many brain areas interconnecting. Whether you think of it as hierarchical or parallel processing, there are identified,

> individual neurons in a network. In the primary visual cortex, or any cortex, we like to think anyway that we know what kind of neurons are in layer one, layer two, where the neurons in layer three project to, where layer four gets their input, that is, the wiring diagram. As I understand current brain analogies in computer science models, one doesn't need to know about the specificities of those layers, because back propagation doesn't require that we understand the details of what's between input and output. But in the brain, I think it does require an understanding of these intricacies to model how an actual brain processes input. As well, for the internal connections of a given brain region, there are external inputs from other brain regions that might be getting completely different information about the world (hormonal, sensory, et cetera). (Interview, May 2022)

We might recall that Hinton's narrative differentiates neurons exclusively on the basis on relative weights; all neurons are in a sense commensurable. Einstein's account, in contrast, indicates different classes of neurons (based on location, which dictates function) and her comments above regarding the endocrine system suggest that modulators of intelligence are diffused throughout the nervous system and difficult to "address" in a fixed way.[17] While the intersections of neuroscientific and electrical engineering imaginaries are evident in the figure of the "wiring diagram," for Einstein it is the methodological implications of the computational claims that are most troubling. She admits to her own investments in the laboratory methods of neuroscience, with its goals of "taking things apart to see what they do," and confesses that in her research she is less disturbed if there is a brain change that doesn't show up behaviorally than she is if there is a behavioral change for which there is no observable brain change.[18] When asked how she accounts for the strength of this commitment she explains:

> I believe behavior is organized by brain circuits. What might ultimately result in behavioral changes might mani-

> fest early in the brain. While there may be a brain change that isn't measurable or for which the measure hasn't been discerned—and therefore, we don't "see" it—if there is a measurable brain change for which a behavior hasn't yet been observed, I'm not surprised. It takes a lot of neurons to organize a behavior. So, some neurons may have changed but not enough yet to yield a behavioral change. (Interview, January 2023)

For Einstein, experimentation is not a mechanistic or reductionist project because she sees the neuron as inseparable from its relations, both within and beyond the body. Methodologically, this poses the ongoing challenge of holding together what she characterizes as "a big world and a minute world" (Interview, May 2022). The former is the person/body/world involved in studies of sex, gender, and women's health, while the latter is the microscopic world of the laboratory sciences. The "immeasurable results"[19] of qualitative research and the measurement systems of experimental science are often difficult to connect: In Einstein's work on cases of female genital cutting, self-assessments of chronic pain using standard indices, physiological indicators, and reported experience from more extended qualitative interviews fail to align in any simple way (Perović et al. 2120), and premenstrual syndrome shows measures of hormone levels that do not correlate with mood (Romans et al. 2012). When the connections do appear, she observes, research interlocutors' familiarity with received narratives of brain–body relations, and their attunement to popular framings of the problems, further complicate the project of determining cause and effect (Interview, May 2022).

At the close of our conversation, Einstein comments that she often thinks about the standard scientistic drive to discover something generalizable and wonders how this can be done without taking modulating systems like the endocrine (or immune, etc.) system and context into account:

> I don't think that the general scientific goal of producing something "fundamental" is going to work very well for

> real-world biological systems. In real biology, difference and variation are what is "fundamental." To find the "true" we need to restrict our claims to the exact conditions/organisms/modulated state under which the study was carried out. If someone homogenizes difference and glosses over variation, they will lose what makes these systems tick. I think we need to be modest in the face of the brain. We can learn something about CA1 pyramidal neurons in aging female Sprague Dawley rats that have had their ovaries removed, but this doesn't really tell us about even the same neuronal type in another species of rodent and certainly not in humans. I tell my students that what we are learning and reporting on is a particular phenomenon, in this animal (human or nonhuman), at this age, in a given context. Perhaps to really know something true is to restrict our claims, and then begin to compare those claims with others. (Interview, May 2022).

The sense of the fundamental that Einstein resists here is one that works to delimit the insides and outsides of the research object—specifically, in the case of computational neuroscience, the brain as neural network—by rendering other systems as epiphenomenal and so extraneous.[20] For Einstein, in contrast, the biological brain is taken as inseparable from the body-in-the-world, and a neuroscience capable of fundamental insight requires methods that expand to incorporate their object's constitutive relations. Rather than containing her research object, in other words, she is committed to reopening and reconfiguring its boundaries as her understanding of it deepens.

## What We Might Learn When the Neural Network Imaginary Encounters Its Limits

Lepage-Richer (2021, 200) adopts the trope of "adversarial epistemology" to argue that knowledge claims for neural networks are "historically contingent on a larger techno-epistemic imaginary which naturalizes an understanding of knowledge as the product of sustained efforts to resist, counter, and overcome an assumed adversary." In the case of computational neural networks, the adversary is not only a brain that hides its secrets but also competing propositions for how those secrets might be disclosed. It might seem ironic that Marcus and others (see Olazaran 1996) attribute the strength of commitment on the part of neural network proponents less to either data or logic than to historic lines of struggle and animus within the computing community, beginning in the 1960s when Marvin Minsky and Seymour Papert drew up their critique of Frank Rosenblatt's "perceptron" (Marcus 2022). The reemergence of neural networks in the 1980s was framed explicitly as an irreconcilable alternative to the symbolic approach.[21] While the symbolic approach was characterized as having reached a dead end, resulting in the so-called AI Winter of the 1980s, Marcus now proposes a new "hybrid" way forward, identified as a "neurosymbolic" approach (Marcus 2022).[22]

In other words, as the capacities of so-called deep learning approaches reveal their limits, there is a return in computational neuroscience to the idea that intelligence requires the manipulation of symbols.[23] Marcus (2022) characterizes symbol manipulation as involving two essential ingredients: "sets of symbols (essentially just patterns that stand for things) to represent information, and processing (manipulating) those symbols in a specific way, using something like algebra (or logic, or computer programs) to operate over those symbols." Once the brain is understood to be involved in symbol manipulation, and symbols translated as code (strings of binary digits or bits), neural processes read as symbol manipulation are ready-made for computational operations. As Marcus explains: "Symbols offer a principled mechanism for extrapolation: lawful, algebraic procedures that can be applied universally, independently of any similarity to known examples. They are (at least for now) still the best way to handcraft knowledge, and to deal robustly with abstractions in novel situations" (2022).[24]

Deep learning approaches hoped for the “emergence” of intelligence given sufficient data along with sophisticated techniques for the detection of potentially meaningful correlations. Yet in their fundamental assumptions and commitments, deep-learning and symbolic approaches share more than they offer up in the way of alternatives. While one relies on statistical analysis and the other on the encoding of algorithms that determine computational operations (so-called rules), both have already translated cognition into a problem of computation before the research begins. Whether a product of stochastic processes or “abstract reasoning,” comprehension is to be achieved through operations enabled by the brain’s capacity to “recognize” and translate input from an externalized world into manipulable numbers, translated in turn into appropriate output.

The brain/mind as an abstract reasoning machine, on this logic, then needs to be put into interaction with a “real world” understood as outside and separate from it. For biology, the connections of brain/body/world are made through relational processes intrinsic to organisms/environments, while for computational neuroscience these are interfaces to be designed. The now well-developed critique of this form of nature/culture dualism is too extensive to rehearse here.[25] But crucial for the purposes of this essay is the premise that there is an inseparable relation between cultural/historical articulations of the real and the real worlds that we as humans inhabit. The delineation of brains and computational systems, and the articulation of sameness and/or difference between them, is not an innocent matter of objective observation but rather a project of worlding[26] in which all of us engaged in the discussion are implicated. And the stakes are not confined to the laboratory but have political, economic, and material consequences for how we draw the boundaries of the human and of other/more-than-human relations.

The moral of these stories that I hope to draw out concerns the limits of a commitment to containment and closure in both theory and method. I have suggested that the commitment to cognitivism, *inter alia,* sustains the closed-world logics of laboratory computational sciences. That commitment seems to leave practitioners with little recourse, when encountering the theoretical and methodological limits of their practice, other than a return that promises a new and salutary synthesis. The alternative, I have suggested, is something closer to the critical technical practice envisioned by Agre (1997, 23) when he wrote:

> Instead of seeking foundations it would embrace the impossibility of foundations, guiding itself by a continually unfolding awareness of its own workings as a historically specific practice. It would make further inquiry into the practice . . . an integral part of the practice itself. It would accept that this reflexive inquiry places all of its concepts and methods at risk. And it would regard this risk positively, not as a threat to rationality but as the promise of better ways of doing things.

As the work of Gillian Einstein makes evident, the fulfillment of Agre's call requires a tolerance for some mess and incompleteness in one's knowledge-making practices and humility regarding one's knowledge claims. In the hands of a critical practitioner, encounters with the contingency and partiality of knowing are taken not as a sign of a failure that needs to be hidden but of the irremediable openness of worldly relations. Those relations involve modes of learning that are deep, not in the sense of the multiplication and ingenious manipulation of homogeneous arrays of numbers but through their implication in practices of ongoing and heterogeneous world-making.

## Notes

I am deeply grateful to Gillian Einstein not only for her feminist technoscientific practice but also for her generous engagement in the development of this chapter. Thanks also to Théo Lepage-Richer and Ranjodh Singh Dhaliwal for the truly generative conversations that led to this book, and to Andrew Clement and Ben Gansky for critical and sympathetic readings of an earlier draft.

1. For indicative works see Dreyfus 1992; Goodwin 1994, 2017; Hutchins 1995; Lakoff and Johnson 1999; Lave 1988; Lynch 1993; Myers 2015; Suchman 2007b.
2. This could be seen as an extension of the strategy adopted by early twentieth-century neuroscientists as described by Star (1992, 213), where "difficulties that could not easily be addressed by some physical or medical model were relegated to 'mind'-related lines of work, such as psychiatry and psychology." The cognitivist project of computational neural networks is to render mind as a technology, relegating the specificities of the brain to the realm of neuroscience while maintaining the position of bodies and worlds as outside its bounds.
3. On the historical connections between closed worlds and cognitivism in the context of geopolitics and computing see Edwards 1996.
4. Brosnan and Michael note the call for a "critical neuroscience" that, with little reference to practice-oriented scholarship in other fields, recovers bodies and worlds through reference to cognate moves within the cognitive sciences, beginning with Varela et al. (1991). Promoting this call, Slaby and Gallagher (2015) posit the existence of what they name "cognitive institutions," of which they propose science in general, and neuroscience in particular, as exemplars. Such institutions, "through various practices and rules, shape our cognitive activity so as to constitute a certain type of knowledge, packaged with relevant skills and techniques" (2015, 35).
5. Or at least of the mainstream of neuroscientific investigation: this is not to say that it couldn't be otherwise, as will be suggested in the second half of this chapter.
6. In biochemistry to denature something is to "destroy the characteristic properties of (a protein or other biological macromolecule) by heat, acidity, or other effect which disrupts its molecular conformation" (*Oxford Dictionary of English).* As in biochemistry, denaturing as a process of decontextualization is unstable, subject to disruption by interactions that can't be contained.
7. https://www.etymonline.com/word/network
8. For an eloquent exposition of the differences that matter between calculation, as a form of reckoning, and judgment see Smith 2019.
9. https://www.cs.toronto.edu/~hinton/. I return to the figure of learning below.
10. More specifically, it is the programmer who defines the system's "objective functions," which in turn determine the relative "correctness" of its outputs, where both are constrained by what a computational system can do. Hinton explains to Thompson that growing dissatisfaction with the labeled data required for back propagation has led to a greater commitment to novel approaches to so-called unsupervised learning. Rather than explicit classification of input, unsupervised learning is the detection of correlations in computationally legible inputs (for example pixels, in the case of computer vision) translated mathematically as "feature detectors," which in turn become the input data for subsequent layers of the network until an effective "data model" is generated. While effectiveness in this context still means that system output is aligned with results assessed as meaningful by associated humans, the generation of that output is further automated and less reliant on human labor.

11 There is of course no inherent contradiction between inspiration and model, as these could easily work in a complementary fashion. The issue is one of accountability, specifically the way in which reversion to the status of inspiration operates here as a hedge on the stronger claim to be engaged in modeling.

12 Ben Gansky (personal communication) points out that Marcus's "prior knowledge" deficit, supposedly solvable through further encoding of abstract concepts, posits a free-floating corpus of knowledge retrievable on demand. Extensively critiqued within feminist theory, this conceptualization ignores realities of learning and intelligence as always specifically situated in lived experience and positionality within a sociotechnical landscape. For a critical examination of the premise that knowledge can be represented as a corpus of commonsense knowledge see Adam 1998.

13 Traveling outside of the laboratory, neuroscientific technologies of imaging and techniques of diagnosis have inspired expansive therapeutic projects, not least in partnership with pharmaceutical industries (see for example Dumit 2004).

14 For a fuller biography see https://einsteinlab.ca/about-us/our-lab/gillian-einstein/.

15 This argument is developed further in Brown et al. 2022.

16 Author interview with Gillian Einstein, May 26, 2022; follow up January 29, 2023. Unpublished. Further references will be in the text.

17 I'm grateful to Ben Gansky (personal communication) for this point. For an argument regarding addressability as a core requirement for all approaches to computing see Dhaliwal 2022.

18 She adds with a laugh, "I only tell that to my best friends" (Interview, May 2022).

19 *Immeasurable Results* is the title of a painting by Lynn Randolph, which provides the frontispiece for Haraway 1997.

20 An analogous case might be theories of language that position context as complicating rather than constitutive of communication. See discussion in Suchman 2007a, chapter 7.

21 As a notable exception, Marcus (2022) cites Hinton's 1990 *Connectionist Symbol Processing*, as "explicitly aimed to bridge the two worlds of deep learning and symbol manipulation"; a project that, Marcus observes, Hinton subsequently abandoned. "When deep learning reemerged [after another brief winter in the early 2000s] in 2012, it was with a kind of take-no-prisoners attitude that has characterized most of the last decade."

22 Yann LeCun, similarly, has recently advocated a "bold new vision" for AI involving a return to a "cognitive architecture" that includes symbolic reasoning, planning, and "common sense" (Heikkilä and Heaven 2022).

23 Marcus (2022) cites as a turning point in overoptimism regarding the power of neural networks the 2021 NetHack "challenge," hosted at NeurIPS 2021 as a partnership between Facebook (now Meta), AI Research, AI Crowd, Oxford, UCL, and NYU, and supported by sponsors Meta AI, and DeepMind. The most recent within the tradition of closed, game-world competitions, the challenge was won by Team AutoAscend with a non–neural net, symbol manipulation–based approach. https://nethackchallenge.com/report.html. Accessed May 26, 2023.

24 Arguably neural networks are already also symbol-manipulating systems, insofar as they rely on curated databases as their training materials.

25 Primary references in feminist theory include Barad 2007; Butler 1993; Haraway 1989, 1991, 1997; see also Latour 1993.

26 "Worlding" is a term introduced within contemporary anthropology to emphasize the ongoing discursive and material practices through which the (always relational) entities that comprise specific cultural and historical realities are enacted, as well as the limits to translation among them. See de la Cadena and Blaser 2018.

## References


Adam, Alison. 1998. *Artificial Knowing: Gender and the Thinking Machine.* New York: Routledge.

Agre, Philip. 1997. *Computation and Human Experience.* New York: Cambridge University Press.

Barad, Karen. 2007. *Meeting the Universe Halfway: Quantum Physics and the Entanglement of Matter and Meaning.* Durham, N.C.: Duke University Press.

Bender, Emily, Timnit Gebru, Angelina McMillan-Major, and Shmargaret Shmitchell. 2021. "On the Dangers of Stochastic Parrots: Can Language Models Be Too Big?" *FAccT' 21.* Accessed May 26, 2023. https://dl.acm.org/doi/10.1145/3442188.3445922.

Brosnan, Caragh, and Mike Michael. 2014. "Enacting the 'Neuro' in Practice: Translational Research, Adhesion and the Promise of Porosity." *Social Studies of Science* 44, no. 5: 680–700.

Brown, A., L. Karkaby, M. Perovic, R. Shafi, and G. Einstein. 2022. "Sex and Gender Science: The World Writes on the Body." In Sex Differences in Brain Function and Dysfunction, ed. C. Gibson and L. Galea, 3-25. Berlin: Springer Cham.

Brown, Nick, and Mike Michael. 2003. "A Sociology of Expectations: Retrospecting Prospects and Prospecting Retrospects." *Technology Analysis & Strategic Management* 15, no. 1: 3–18.

Butler, Judith. 1993. *Bodies That Matter: On the Discursive Limits of "Sex."* New York: Routledge.

de la Cadena, Marisol, and Mario Blaser, eds. 2018. *A World of Many Worlds.* Durham, N.C.: Duke University Press.

Dennett, Daniel. 2013. *Intuition Pumps and Other Tools for Thinking.* New York: Norton.

Dhaliwal, Ranjodh Singh. 2022. "On Addressability, or What Even Is Computing?" *Critical Inquiry* 49, no. 1: 1–27.

Dreyfus, Hubert. 1992. *What Computers Still Can't Do.* Cambridge, Mass.: MIT Press.

Dumit, Joe. 2004. *Picturing Personhood: Brain Scans and Biomedical Identity.* Princeton, N.J.: Princeton University Press.

Edwards, Paul. 1996. *The Closed World: Computers and the Politics of Discourse in Cold War America.* Cambridge, Mass: MIT Press.

Einstein, Gillian. 2012. "Situated Neuroscience: Exploring Biologies of Diversity." In

*Neurofeminism,* New Directions in Philosophy and Cognitive Science, ed. R. Bluhm, A. J. Jacobson, and H. L. Maibom, 145–74. London: Palgrave Macmillan.

Fausto-Sterling, Ann. 2012. *Sex/Gender: Biology in a Social World.* New York: Routledge.

Goodwin, Charles. 1994. "Professional Vision." *American Anthropologist* 96, no. 3: 606–33.

Goodwin, Charles. 2017. *Co-Operative Action.* Cambridge: Cambridge University Press.

Hall, Stuart. 2003. "Marx's Notes on Method: A 'Reading' of the 1857 Introduction." *Cultural Studies* 17, no. 2: 113–49.

Halpern, Orit. 2022. "The Future Will Not Be Calculated: Neural Nets, Neoliberalism, and Reactionary Politics." *Critical Inquiry* 48, no. 2: 334–59.

Haraway, Donna. 1989. *Primate Visions: Gender, Race, and Nature in the World of Modern Science.* New York: Routledge.

Haraway, Donna. 1991. *Simians, Cyborgs, and Women: The Reinvention of Nature.* New York: Routledge

Haraway, Donna. 1997. *Modest_Witness@Second_Millenium.FemaleMan_Meets_OncoMouse™: Feminism and Technoscience.* New York: Routledge.

Heikkilä, Melissa, and Will Douglas Heaven. 2022. "Yann LeCun Has a Bold New Vision for the Future of AI." *MIT Technology Review,* June 24. Accessed May 26, 2023. https://www.technologyreview.com/2022/06/24/1054817/yann-lecun-bold-new-vision-future-ai-deep-learning-meta/.

Hutchins, Edwin. 1995. *Cognition in the Wild.* Cambridge, Mass.: MIT Press.

Inioluwa, Deborah Raji, Emily Bender, Amandalynne Paullada, Emily Denton, and Alex Hanna. 2021. "AI and the Everything in the Whole Wide World Benchmark." *35th Conference on Neural Information Processing Systems (NeurIPS 2021).* https://arxiv.org/pdf/2111.15366.pdf.

Lakoff, George, and Mark Johnson. 1999. *Philosophy in the Flesh: The Embodied Mind and Its Challenge to Western Thought.* New York: Basic Books.

Latour, Bruno. 1993. *We Have Never Been Modern.* Cambridge, Mass.: Harvard University Press.

Lave, Jean. 1988. *Cognition in Practice: Mind, Mathematics, and Culture in Everyday Life.* Cambridge: Cambridge University Press.

Lave, Jean. 2011. *Apprenticeship in Critical Ethnographic Practice.* Chicago: University of Chicago Press.

Lepage-Richer, Théo. 2021. "Adversariality in Machine Learning Systems: On Neural Networks and the Limits of Knowledge." In *The Cultural Life of Machine Learning: An Incursion into Critical AI Studies,* ed Jonathan Roberge and Michael Castelle, 197–225. London: Palgrave.

Lynch, Michael. 1988. "Sacrifice and the Transformation of the Animal Body into a Scientific Object: Laboratory Culture and Ritual Practice in the Neurosciences." *Social Studies of Science* 18, no. 2: 265–89.

Lynch, Michael. 1993. *Scientific Practice and Ordinary Action: Ethnomethodology and Social Studies of Science.* New York: Cambridge University Press.

Marcus, Gary. 2018. "Deep Learning: A Critical Appraisal." *arXiv preprint arXiv: 1801.00631*

Marcus, Gary. 2022. "Deep Learning Is Hitting a Wall." *Nautilus.* March 10. Accessed October 2022. https://nautil.us/deep-learning-is-hitting-a-wall-238440/.

Marcus, Gary, Adam Marblestone, and Thomas Dean. 2014. "The Atoms of Neural Computation." *Science,* 346 (6209), 551–52.

Martin, Emily. 1991. "The Egg and the Sperm: How Science Has Constructed a Romance Based on Stereotypically Male-Female Roles." *Signs: Journal of Women in Culture and Society* 16, no. 3: 485–501.

Martin, Emily. 2004. "Talking Back to Neuro-reductionism." In *Cultural Bodies: Ethnography and Theory,* ed. H. Thomas and J. Ahmed, 190–211. New York: Wiley & Sons.

McCarthy, John, Marvin Minsky, Nathaniel Rochester, and Claude Shannon. 1955. "A Proposal for the Dartmouth Summer Research Project on Artificial Intelligence." Accessed May 26, 2023. http://raysolomonoff.com/dartmouth/boxa/dart564props.pdf.

Myers, Natasha. 2015. *Rendering Life Molecular: Models, Modelers, and Excitable Matter* Durham, N.C.: Duke University Press.

Olazaran, Mikel. 1996. "A Sociological Study of the Official History of the Perceptrons Controversy." *Social Studies of Science* 26, no. 3: 611–59.

Perović, Mateja, Danielle Jacobson, Emily Glazer, Caroline Pukall, and Gillian Einstein. 2021. "Are You in Pain if You Say You Are Not? Accounts of Pain in Somali-Canadian Women with Female Genital Cutting." *PAIN: The Journal for the International Association for the Study of Pain* 162, no. 4 (April 2021): 1144–52.

Romans, Sarah, Rose Clarkson, Gillian Einstein, Mateja Perović, and Donna Stewart. 2012. "Mood and the Menstrual Cycle: A Review of Prospective Data Studies." *Gender Medicine* 9, no. 5: 361–84.

Schull, Natasha, and Caitlin Zaloom. 2011. "The Shortsighted Brain: Neuroeconomics and the Governance of Choice in Time." *Social Studies of Science* 41, no. 4: 515–38.

Slaby, J. and S. Gallagher. 2015. "Critical Neuroscience and Socially Extended Minds." *Theory, Culture & Society* 32, no. 1: 33–59.

Smith, Brian Cantwell 2019. *The Promise of Artificial Intelligence: Reckoning and Judgment.* Cambridge, Mass.: MIT Press.

Star, Susan Leigh. 1992. "The Skin, the Skull, and the Self: Toward a Sociology of the Brain." In *So Human a Brain,* ed. Anne Harrington, 204–28. Boston: Birkhäuser.

Suchman, Lucy. 2007a. *Human-Machine Reconfigurations: Plans and Situated Actions,* rev. ed. New York: Cambridge.

Suchman, Lucy. 2007b. "Feminist STS and the Sciences of the Artificial." In *The Handbook of Science and Technology Studies,* 3rd ed., ed. E. Hackett, O. Amsterdamska, M. Lynch, and J. Wajcman, 139–63. Cambridge, Mass.: MIT Press.

Thompson, Nicholas. 2019. "An AI Pioneer Explains the Evolution of Neural Networks." *Wired,* May 13. Accessed May 26, 2023. https://www.wired.com/story/ai-pioneer-explains-evolution-neural-networks/.

Varela, Francisco, Thompson, Evan, and Eleanor Rosch. 1991. *The Embodied Mind: Cognitive Science and Human Experience.* Cambridge, Mass.: MIT.

Vidal, Fernando. 2009. Brainhood, Anthropological Figure of Modernity. *History of the Human Sciences* 22, no. 1: 5–36.

Vidal, Fernando, and Franciso Ortega. 2011. "Approaching the Neurocultural Spectrum: An Introduction." In *Neurocultures: Glimpses into an Expanding Universe,* ed, F. Ortega and F. Vidal, 7–27. Frankfurt am Main: Peter Lang.